Whitetails

A PHOTOGRAPHIC JOURNEY
THROUGH THE SEASONS

Text and photography by

Charles J. Alsheimer
www.CharlesAlsheimer.com

Published by

Krause Publications, a division of F+W Media, Inc.
700 East State Street • Iola, WI 54990-0001
715-445-2214 • 888-457-2873
www.krausebooks.com

To order books or other products call toll-free 1-800-258-0929
or visit us online at www.krausebooks.com or www.Shop.Collect.com

ISBN-13: 978-1-4402-2866-7
ISBN-10: 1-4402-2866-3

Cover Design by Al West
Designed by Dusty Reid
Edited by Corrina Peterson

Printed in the United States of America

Dedication

TO THE GOD OF THE HOLY BIBLE
Maker of Heaven and Earth

Contents

Foreword

You are about to begin a special year-long journey into the secret, little-known world of the majestic white-tailed deer, one of God's most beautiful creatures. You will feel like you are there as you follow six deer named Bucky, Skipper, Buttercup, Daisy, Buttons and Princess.

Starting in April, with the first breath of spring, the book reveals seldom-seen and little-understood activities in the life of the whitetail. You will watch as Daisy, a mother doe, drives off her fawns from the previous year before delivering another set of twins. Then you will follow as Skipper, the male fawn, searches out a new territory and runs into Bucky, the monarch of the forest. You will feel like Skipper feels as he becomes part of a summer bachelor group and will be amazed how he handles his first rut in fall.

You will also see fawns born, the complex communication system of does and bucks revealed, and huge snowflakes pile up on the backs of whitetails as they ride out a winter storm. You will observe what happens as deer encounter turkeys, coyotes, squirrels, blue jays, farm combines, hunting seasons and other situations. And you will see why many people believe deer are one of the most intelligent animals.

Without a doubt, this book by award-winning writer and photographer Charles J. Alsheimer is like no other piece of whitetail literature, for it accurately takes the reader into the lives of six white-tailed deer for an ex-

tended time. Written in an easy-flowing style, the book lets everyone from a middle-school youngster to a college graduate enjoy what's revealed about these deer.

In addition, beautiful photographs are found throughout the book, perfectly illustrating the text. For example, when Alsheimer describes how a deer learns to stand on its hind feet to snag an apple, a picture shows the buck grabbing an apple from the tree. These photos put great impact and emotion into each page and bring out the passion the author has for deer.

In fact, without Alsheimer's passion and love for the whitetail, it would not have been possible to have the motivation to complete a masterpiece such as this. Alsheimer has photographed deer since 1971. Since 1995, he has spent extensive time with whitetails throughout the year. While talking with him recently, I calculated that during the past 15 years he has spent more than seven years of 40-hour work weeks photographing and studying deer.

Charlie and I have been friends for many years. Although his previous six whitetail books carry the highest rating possible from readers, I believe this is his best book. And, interestingly, though it's not a hunting book, what you learn within these pages will make you a better hunter.

Brad Herndon
Brownstown, Ind.

Acknowledgements

In September 1979, I embarked on a journey to chase a dream of becoming an outdoor writer and nature photographer. At the time, many said I was living in a fantasy world because what I wanted to do was a crap shoot or pipe dream. Fortunately, some in my inner circle thought I could pull it off. On my office wall, I have a wooden sign with the quote, "Live Your Dreams." Well, for 30 years, I've been able to live the dream.

What God has allowed me to accomplish required some talent, but most of the credit for my success goes to those who made my dream possible. I've been blessed with the presence of some very special, caring people. Whenever you try to acknowledge those who have helped, you run the risk of leaving someone out. Hopefully, I will not do that, but if I do, please forgive me.

Charles H. and Eleanor Alsheimer: Though you are gone, I want to thank you for introducing me to the wonders of nature. Dad, you sacrificed a lot by taking me with you on your hunts. And Mom, thank you for encouraging me to be all I could be.

Carla: To the love of my life, I say thanks. You've made it possible for me to fly. You gave me my wings, kept me on course and encouraged me along the way. Thank you for being a wonderful wife for more than 38 years and my best friend.

Aaron: Aside from being an incredible photo model, you're the greatest son any man could hope to have. For the past 32 years, we've had a storybook relationship as father and son. Together, we've climbed to the top of the Rockies, bushwhacked our way through the Alaskan tundra, canoed the Everglades to the Atlantic Ocean, explored many of the wild haunts in North America and spent untold hours in the whitetail forest. It's been a special trip. Thanks for loving me, thanks for being there, and thanks for the memories.

Paul Daniels: Thanks to one of my best friends in the world. You've always been willing to do anything I've needed, from modeling to caring for the deer operation. I love ya, man!

Terry Rice: Thanks for all the early-morning modeling sessions. You're great at what you do, and your friendship means a lot to me.

Dick Snavely: Thank you for being a very special mentor. I don't know where I'd be today had you not introduced me to Jesus Christ in 1971. Thanks for always being there to help and mold me into the person I am.

Haas Hargrave: Thanks for nudging me in Summer 1979 to chase this dream. Your wise counsel helped birth a dream that became a reality. You've been an incredible boss, mentor and friend.

Jack Brauer and Al Hofacker: You launched Deer and Deer Hunting magazine, which no one thought was possible. Then, you took a chance on an obscure farm guy from western New York who had a big dream. You gave me my first real break in this business and coached and nurtured me through the early years. It's doubtful I'd be where I am today without you. Thanks for who you are, and thanks for what you did for me.

Pat Durkin and Dan Schmidt: Of all the editors I've worked with in the outdoor world the past 30 years, you're the best. During your time in the Deer and Deer Hunting editor's chair, you gave me a lot of encouragement and leeway. You helped me grow. I'm forever indebted for all you have done for me. I count our friendship as being very special.

Dick Bernier: Thank you for your research help and big-woods tracking insight. But most of all, thank you for your friendship. You are the best deer tracker in the business. Stay on the track!

Brad Herndon: Thank you for nudging me into digital photography. Your guidance in the digital world has been enlightening and a godsend. You are a special friend.

Bob and Alma Avery: Though Bob is gone, I want to thank you for loving me, feeding me, adopting me into your family and letting me photograph whitetails on your incredible property.

Jim, Charlie, Jack, Aaron, Paul, Whitey, Spook, Dodger, Bambi, Bucky, Yankee, Carla, Susie, Tiny and Buttercup: Without you, I wouldn't have learned what I know about whitetails. Collectively, you have taught me more than the wildlife biologists or scientific journals combined. Thanks for allowing me a window into the whitetail's hidden world.

Last but most important, I want to thank Jesus Christ for the gift of life. To some, the mention of His name is a turn-off. To others, He is looked at as a crutch. But to me, He is the reason for living, hope and the success I've had. I owe what I have today to the grace of God. It's that simple. I serve a great and loving God, and without Him, none of this would be possible.

About the Author

Charles Alsheimer is an award-winning lecturer, outdoor writer, nature photographer and whitetail consultant from Bath, N.Y. Alsheimer was born and raised on a farm, and has devoted his life to photographing, writing and lecturing about the wonders of God's creation. His specialty — as a writer and photographer — is the white-tailed deer.

He is the senior contributing editor for Deer and Deer Hunting magazine and host of its national television show, Deer & Deer Hunting TV, which airs on the Versus Network. He is also a contributing editor for Whitetail News.

For more than 30 years, Alsheimer's work has taken him across North America. His photography has won numerous state and national contests, and his articles and photographs have appeared in nearly every major outdoor publication, including Outdoor Life, Field & Stream, Sports Afield, Harris publications and Deer and Deer Hunting. In addition, he has written six popular books on the whitetail and co-authored a seventh. Alsheimer also owns and operates a white-tailed deer research facility and provides consulting services to various segments of the whitetail industry.

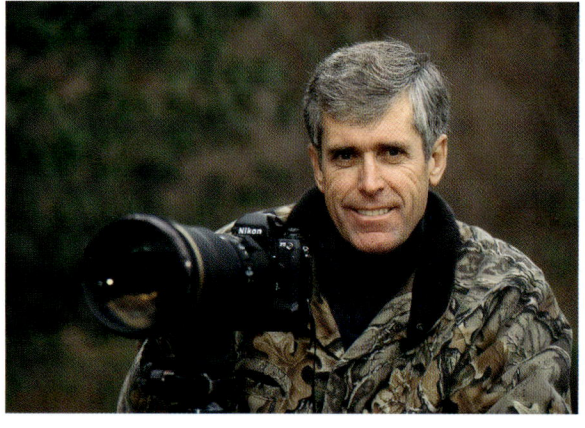

In a national poll conducted in January 2000 by Deer and Deer Hunting, Alsheimer was honored as one of deer hunting's top five inspirational leaders of the past century. The ballot included the names of nearly 60 scientists, manufacturers, politicians, celebrities, communicators and hunters. These people were selected for their efforts to enhance America's understanding of the white-tailed deer and promote the sport of deer hunting by establishing national forests, developing camouflage patterns and writing popular books. When the results of the survey were compiled, Alsheimer ranked third behind bowhunter Fred Bear and conservationist Aldo Leopold. President Theodore Roosevelt and outdoor apparel designer Bill Jordan finished fourth and fifth, respectively. This honor illustrates the respect Alsheimer has among America's deer hunters.

Alsheimer lives with his wife on their farm in rural upstate New York. Additional information about him and his work can be found on his website, www. CharlesAlsheimer.com.

Introduction

I've wanted to write this book for a very long time. I'd hate to think what my life would have been like without the white-tailed deer. It was what introduced me to nature. As a little boy, the graceful figure of a mature buck running across a plowed field on our farm was what lit my fire — a fire that has kept me heading back to the woods for more than 50 years.

In my wildest dreams, I never envisioned having a career in the outdoor field. I dreamed some pretty big dreams growing up on a rural western New York farm in the 1950s and '60s. Back then, baseball and hunting dominated my life, and my dream was to be the next Mickey Mantle or Jack O'Connor. They were iconic figures in their respective professions, Mantle in baseball and O'Connor in the hunting world. They were my heroes, and I wanted to live their lives. Unfortunately, almost every time I shared my dream with someone, they said I needed to get real because there was no chance of that happening to someone from where I lived.

Despite those early reality checks, I kept chasing my dreams, no matter how unrealistic they seemed to others. Along the way, I became a pretty fair ball player, but I became a better hunter. When I was 20, I realized the baseball dreams I had as a teen-ager were coming to an end. I wasn't good enough to play baseball professionally, so my focus shifted to hunting.

In 1969, while serving in the U.S. Air Force, I was deployed to Vietnam for 14 months. Despite some harrowing experiences, my time there was life-changing. For starters, life in a war zone made me appreciate America like never before. While there, my boss, a tech sergeant from Iowa, introduced me to photography. He knew of my passion for hunting and encouraged me to use my hunting skills to photograph nature when I returned to the states. So, I bought cameras and telephoto lenses, and began photographing white-tailed deer and nature after my discharge from the military. Little did I know that the 35mm cameras would be a foundational block to help me chase my dream of making a living as an outdoor writer and nature photographer.

After discharge from the military, I returned to college and married. When I graduated from college, I took a job in sales and marketing with a large furniture manufacturer. During the next seven years, I received an incredible business education working for Haas Hargrave, the company's vice president of sales. Haas had been a bomber pilot in World War II, an All-American college football player and a college football coach before his business career. He was also a passionate deer hunter. During the time I worked for him, we talked a lot about deer hunting. He also taught me invaluable business lessons that would serve me well the rest of my life.

One morning, Haas poked his head into my office and told me he needed to talk to me. When I walked into his office, he asked me to close the door and sit down. As I sat in front of his desk, he leaned over, took a magazine out of his briefcase and handed it to me. It was a copy of Deer and Deer Hunting's magazine called The Stump Sitters. He said he had picked it up a day earlier in Milwaukee and noticed the magazine was looking for writers. He told me I should try to write for them. I just stared at him, not knowing quite what to say.

Then he said, "Everyone needs to be a pioneer at least once in his life." What he meant was that we need to step out of our comfort zone and pursue what we love because life is short. Having worked together for five years, Haas couldn't see me spending my working career in corporate sales. He encouraged me to work for myself.

He asked if I had thought of being a full-time outdoor writer and nature photographer. My response was something like, "Yes, but I'm not sure I can make a living at it." His comeback was, "You won't know unless you try," and he suggested I follow up on the lead in the magazine to see if the editors might give me a shot.

With Haas' encouragement, I queried the magazine with two article ideas. To my surprise, they liked and purchased both. That led to other articles, and before long, I began writing an outdoor column for our local newspaper. The journey was heating up.

In August 1979, with butterflies in my stomach, I resigned my sales and marketing position and said goodbye to the manufacturing world. On the eve of Labor Day weekend, I traded my coat and tie for hunting clothes. At the time, my dad thought I had lost my mind because I was leaving a good job to pursue a career filled with many unknowns. To say those early days were scary would be an understatement.

Cast

Bucky: A majestic 6-year-old buck.
Skipper: A yearling buck trying to find his place in the whitetail world.
Buttercup: An 8-year-old doe, wise beyond her years.
Daisy: Buttercup's yearling daughter.
Buttons: Buttercup's buck fawn, born at the end of May.
Princess: Buttercup's doe fawn, born at the end of May.

For the first three years, I photographed and wrote passionately about the whitetail. Because I poured everything I earned into camera equipment and travel expenses, I didn't make a profit. Fortunately, I had a very supportive wife who had a good-paying job to help pay for my dream. We also had a little boy who was incredible. Aaron loved the outdoors as much as I did and went with me wherever I went. The support my family gave me made it easier to chase my dream.

Others helped make my journey easier. In the early 1980s, I contacted two of the biggest names in the hunting and nature photography world for advice: Erwin Bauer and Lenny Rue. They could have told me to get lost. Instead, they reached out and mentored me. Believe me, you don't make it in this business by yourself, so I'm forever indebted to both for the help and guidance they provided.

However, it's safe to say that my career would never have gotten off the ground had I not committed my life to Christ nearly 40 years ago. I realize that professions of faith are real turn-offs for many. But in my case, my faith in God has been the guiding force in my life. He's let me live in the greatest country on Earth, given me an incredible family, brought individuals into my life who have enriched my days beyond measure and let me have a dream job. A man can't ask for more.

For me, being an American is a great honor. When people ask me to reflect on my life, the lyrics to Aaron Tippin's top-selling country hit Where the Stars and Stripes and the Eagles Fly roll off my lips: "Well, if you ask me where I come from, here's what I tell everyone. I was born by God's dear grace, in an extraordinary place — where the stars and stripes and the eagle flies. It's a big old land with countless dreams. Happiness ain't out of reach, hard work pays off the way it should. Yeah, I've seen enough to know we've got it good, where the stars and stripes and the eagle flies." These lyrics say it all.

More than 30 years have passed since Hargrave challenged me to chase the dream of becoming a full-time outdoor writer and nature photographer. Along the way, I've hunted whitetails and photographed in some of the most amazing country imaginable — and do things my parents never would have imagined. It's been a special trip.

But the beauty of the trip has been this: Though goals might be the No. 1 end game for some, they aren't to me. I've realized that life's journey is the true destination. That's what has made my life so special.

Because of my life's journey, this book has been percolating in my mind for more than 25 years. During this time, I've lived with white-tailed deer. That's a pretty big statement, but it's true. Growing up in a farming family that hunted provided me opportunities to study whitetails at a young age. But from 1979 on, my life with whitetails evolved from merely hunting them to intensely photographing them to raising them to study their behavior.

Unlike the six previous whitetail books I've done, this title delves into the inner life of six white-tailed deer; what they encounter in the wild, how they think and how they survive. The setting is the northeastern United States, which is home to the northern woodland subspecies of whitetail. In these pages, you will see some of the best photos I've taken to illustrate my belief that the whitetail is the greatest animal God created. To complement the photos, I've woven in many behaviors I've witnessed for more than 50 years.

Yes, my life with whitetails has been very special. As you turn these pages, I hope you'll gain a greater insight into this incredible animal. Enjoy the journey.

Charles J. Alsheimer
Bath, NY
May 1, 2010

CHAPTER 1

April

The gurgle of flowing stream water mesmerized Buttercup and Daisy as they lay bedded under a cluster of low-growing hemlock trees feet from the water's edge. It had been hours since they had fed in the pre-dawn darkness. Hungry and sensing nightfall's approach, they rose from their bed and stretched. But before they could take a step, their ears picked up the sound of something approaching. For several seconds, they stared upstream, intently scouring the undergrowth to see what was coming. Neither doe dared to move for fear of being detected.

Daisy was the first to see movement. Two coyotes, with their heads down in a stealth crouching posture, were slowly moving toward the deer. Neither coyote saw the does, but they smelled them. The coyotes knew that if they took their time, were cautious and worked as a team, they would be in for a venison feast by nightfall.

Sensing what was about to occur, Buttercup flared her tail to alert nearby deer of the approaching danger. Before either coyote could spot them, Buttercup and Daisy wheeled and bolted from the hemlock thicket, scattering leaves and breaking branches in their wake. The chase was on.

▲ By April, most bucks have started growing antlers. If stress is low and nutrition high, some bucks can carry their antlers into April, like the buck in this photo.

▲ During April, deer begin to shed their winter fur.

For the next few minutes, the does ran for their lives. Neither looked back, but they knew the coyotes were closing in. As they ran and bounded through the woods, the does felt the toll of the long winter, as they lacked the strength and stamina they'd had the previous fall.

The does left the woods and bounded across a fallow field before stopping to catch their breath and assess the situation. With their tongues hanging out and chests heaving, they saw the coyotes enter the field and run toward them. The deer knew that unless they could put more distance between them and the coyotes, they would soon be overcome with exhaustion.

First Daisy and then Buttercup wheeled and ran into the hardwoods. Fifty yards inside the woods, the forest dropped off into a valley. The going was tougher, as the deer negotiated one deadfall after another. Near the bottom of the hillside, the woods flattened out for a few yards. Buttercup, exhausted from the chase, stopped next to a fallen oak tree. Then she looked uphill to see the horror all whitetails fear. The larger of the coyotes was less than 10 yards away, running straight for her. He was snapping his jaws as he moved in for the kill.

Without hesitating, the big male dog jumped for Buttercup's hind legs, just as she lurched to avoid the attack. He missed her leg's tendon but latched his jaws into her flank, coming away with a mouth full of fur. Adrenalin flooded Buttercup's body. Fearing death's grasp, she thrust herself down the hill toward the spring-swollen river, knowing it held her only chance of survival. Running full speed, with panting coyotes on her heels, she raced through the matted grass at the raging water's edge and catapulted her body into the air. The big doe hit the water at full speed. Immediately, the current pulled her away from the bank toward the center of the river, with only her head bobbing above the fast-moving water.

Quickly, Buttercup flailed her legs in doggy-paddle fashion. When she reached the middle of the river, she spotted Daisy frantically swimming a few yards in front of her. With heads bobbing above the current, the does swam to the other side of the muddy river. Exhausted, the deer used every bit of energy they had to climb up the river's bank. They were safe, because the coyotes had abandoned the chase at the river's edge.

Buttercup and Daisy slowly walked into a tangle of brush, shook the river water from their body and stood motionless for several minutes. The events of the previous hour had been harrowing and exhausting. Though hungry, the does couldn't eat because they were too tired. It was time to recover from what they had endured. Feeling safe, mother and daughter bedded until nightfall.

After darkness arrived, the does rose from their bed and stretched. Daisy began licking her mother's flank where the coyote had ripped the fur from Buttercup's body during the chase. For the next several minutes, the does groomed each other before moving to a nearby field to feed.

As they entered the field, the does saw there were other deer already feeding on newly sprouted spring forage. For the next hour, the does filled their stomachs with tender new growth before bedding. During the night, Buttercup and Daisy fed and bedded several more times. It seemed good to gorge themselves on lush food after going through a winter with little to eat. Twice during the night, the chorus of coyotes howling in the distance put the deer on full alert. It was a reminder of the previous day's events and the need to be vigilant to the threats of their world.

By the time a Northern winter is finished, most whitetails have lost 20 percent of their pre-rut body weight.

Soon after dawn, the warm spring rain stopped. Fog began easing through the forest, riding the draft of the wafting thermals. As the minutes passed, the forest grew brighter. April and spring's rebirth was new to Skipper. The young buck was a couple of months shy of his first birthday and had never seen the things unfolding before him. The smells of spring overloaded his senses, and the behavior of other wildlife seemed to amaze him. His life was one of discovery.

Skipper had bedded at the edge of a meadow a couple of hours before dawn. In the rainy, drippy forest, he had felt alone in the dark. But when daylight arrived, he quickly discovered an abundance of life. In the meadow 50 yards from him was a red fox slowly making its way across the opening. Lying there, with droplets of water landing on his body from overhanging branches, the young buck watched as the fox pounced and caught several mice. In the distance, he heard the gobbles of turkeys greeting the arrival of spring.

▼ During April, a deer's search for food causes them to be active throughout the day.

▼ Many yearling bucks begin dispersing in April. During the process, they often attempt to bond with other bucks.

Whitetail bucks will scent-mark overhanging licking branches throughout the year. This behavior is a whitetail's way of communicating with other deer in their territory.

After a long, hard winter, it's critical for does to find highly nutritious food during the final two months of gestation.

If adequate forage is available, a whitetail will consume up to 10 pounds of food per day.

Soon after the red fox left the meadow, two whitetails emerged from the far side of the field and began feeding. Skipper, feeling social, rose from his bed, stretched and began walking toward the deer. Several days had passed since he'd left his mother. By Buttercup's actions, it was obvious she did not want him around anymore, so he left the forest that had been his home since birth and explored other areas. He had never seen any of this real estate, and none of the deer he was encountering were familiar.

He walked to the two deer and discovered they were bucks, much bigger than him. He tried to socialize, but neither would have anything to do with him. Each time Skipper tried to approach the bucks, the larger deer would pull its ears back, bristle up his hair and walk around him in a stiff-legged manner, threatening to attack him. Skipper knew by the buck's behavior that he should leave. Within seconds, he backed off and left the field.

By midday, the skies had cleared. It was a beautiful day; sunny and windless, though still quite cool. Throughout the day, Skipper wandered, fed, bedded and then wandered some more. He felt like he was on a mission but wasn't sure where he was going. With less than an hour of daylight remaining, the young buck stepped into a winter wheat field. The tender new growth would provide a great meal, so he began to feed. Within minutes, he heard a twig snap in the woods nearby. On alert, Skipper turned his head and focused on the sound's source in the woods. At first, he saw nothing, but after staring intently for a few moments, he picked up the outline of a deer coming toward him.

Skipper didn't take his eyes off the deer as it approached. When the deer got to the field's edge, Skipper knew it was another buck — much bigger than the two he had encountered earlier that day. The big buck, called Bucky, paid little attention to Skipper as he walked toward him along the field's edge. Rather than stepping into the field to feed, the big buck walked to an overhanging tree branch and began rubbing his forehead, eyes and nose on the end of the branch. Several times, he stopped to lick the branch. When done marking the branch with his scent, Bucky spread his legs apart and urinated on the ground under the branch. Bucky had called the area home for more than five years and marked the branch to let every deer know he was alive.

Vital Information

• By April 1, most whitetails will have left their winter habitat and returned to familiar summer range.
• In farm country, this movement can be as little as one to two miles. However, in remote country, the migration might be more than 20 miles.
• During April, pregnant does begin to drive off their buck fawn(s) from the previous year. This is nature's way of preventing inbreeding.
• In Northern regions, whitetails begin shedding their winter coat in early April, which makes them look rather ugly.
• In early April, it's not uncommon for whitetails to weigh 20 percent less than they did before the previous November's rut. Therefore, a whitetail's food intake can be four times greater than it was in March, providing April green-up occurs early in the month.
• When green-up arrives, the protein levels of new-growth forages such as clover and alfalfa can be more than 15 percent, which is vital for does during the final two months of gestation and bucks as they begin antler growth.
• As bucks' antlers start growing, their scent marking behavior increases. One way they do this is by leaving scent from their forehead, preorbital and nasal glands on overhanging branches throughout their area.
• In addition to scent marking, whitetails also communicate through vocalization. One of the most common sounds they make is snorting. An alarm snort is loud and made in response to danger.
• Hoof stomping is an alarm signal done by a buck or doe to alert other deer of imminent danger. This behavior is often made in conjunction with alarm snorts.
• It has been estimated that whitetails can smell up to 10 times better than humans.
• If tracks made by a whitetail standing still are wider than 2½ inches, they were probably made by a mature buck. If such tracks are less than 2 inches wide, they were most likely made by a doe or yearling buck.
• Research has shown that whitetails can probably see colors in the blue and yellow range quite well. It's believed they see red and green colors as shades of gray.

Slowly, Bucky walked toward Skipper. The small buck didn't move. When the big buck approached to within inches of Skipper's head, it stared into the young buck's eyes as if to say, "Hi, I'm the boss around here. Can I ask you to groom my neck and shoulders?" Without taking his eyes off of Bucky, Skipper smelled the big buck's neck and front shoulder and began to lick the monarch. Within seconds, Bucky began to return the favor by licking Skipper. For the next few minutes, the bucks groomed each other's neck, shoulders and forehead. After tiring of the mutual massage, the bucks separated and began feeding until darkness fell.

The wind howled throughout the night, making trees sway back and forth. The sound of branches constantly hitting together made Bucky and Skipper skittish, so they left the woods to find a less threatening location to bed. They settled on an alder swamp not far from where they had met the previous day. Though the wind was still blowing, they didn't have to contend with the noise of the woods.

As darkness eased toward the gray of dawn, the temperature dropped, and it began to rain. An hour after daybreak, the rain changed to snow. Skipper must have thought, "What a difference a day makes. Yesterday, it was cool and sunny, and today it's cold and snowy. Just when I thought winter was over, it tries to make a comeback." For the next two hours, big flakes fell from the sky, carpeting the ground with two inches of wet, fluffy snow. Clearly, winter wasn't finished yet.

By late afternoon, the skies had cleared, and the snow began to melt. Skipper and Bucky rose from their beds in the alder swamp and headed to the wheat field where they had met. By the time they poked across a fallow field and arrived at the wheat field, several deer were already feasting on the winter wheat just beneath the blanket of snow. Neither buck acknowledged the other deer as they began to feed. Every deer in the field was on a mission to recover physically from the harsh winter. Socializing would have to wait.

The night was clear, cold and moonless. The infinite canopy of stars gave the deer enough light to feed. Throughout the night, the wheat field received a lot of attention from the local herd. Bucky and Skipper fed and bedded near each other, though neither made any attempt to socialize. Just before dawn, the bucks fed one last time before disappearing into the forest. May and fawning season were just around the corner.

During May does often fight each other to secure territorial locations to birth their fawns.

CHAPTER 2

May

For the previous three years, Bucky's favorite May destination had been a local farmer's big clover field. Its vast sea of green plants provided most of the area's deer with nourishment to recover from the long, cold winter. The field was in the heart of Bucky's home range, so he fed there daily.

Throughout the night, Bucky fed in the clover field several times. With two hours of darkness remaining, he headed to his favorite daytime bedding site: a long hemlock ridge about 200 yards from the clover field. Although the forest was dark, Bucky quickly traveled to one of his favorite bedding areas, overlooking several fingers of valleys, to get a few hours sleep before daybreak. As dawn approached, he dozed on and off, periodically waking for a few minutes to chew his cud.

As the first hint of a new day arrived, Bucky was awakened from a sound sleep when he heard, "Gobble-gobble-gobble. Gobble-gobble-gobble!" The tom turkey's booming gobbles echoed down the valley. A lovesick gobbler was roosted high in a big hemlock tree less than 50 yards away. Though Bucky had grown accustomed to the sound of gobbles and yelps at all times of day, he knew he wouldn't get more sleep until the bird flew down. Rather than leave, the big buck decided to stay bedded and take in the gobbler's pre-dawn serenade.

Between the turkey's gobbles, the big buck closed his eyes and tried to doze. It was futile. Not only was the tom gobbling every few minutes, but hens were yelping between gobbles. The back-and-forth between the gobbler and hens was nature's symphony, which is played out every spring. When the first rays of sunlight touched the treetops, three hens dropped from their roost and landed within yards of Bucky. He watched intently as they walked toward him, yelping with every step. The largest hen spotted him in his bed. Neither felt threat-

▼ When in bloom, large flowered trilliums are highly preferred by whitetails.

During May, when forage is abundant, deer will feed throughout the day, consuming up to 15 pounds of food.

▲ Just before giving birth, a doe will attempt to drive all deer away from her selected birthing location.

▲ In some cases, a doe will tolerate the presence of a buck in the weeks before fawning. However, that's not the case when fawning is near.

ened by the other, so they looked at each other. Before the hens walked out of sight, the tom gobbled twice and then flew down from his perch, scattering leaves as he landed. The big bird immediately gobbled three times before heading toward the hens. As the tom walked past, it stopped to look Bucky over carefully, slowly cocking his head from side to side. Then, as if to scold Bucky, the bird gobbled twice more. Though the sound was deafening, the big buck never moved. Then the gobbler hurried through the woods to catch up with his hens.

When the turkeys were out of sight, the woods became much quieter. Now and then, Bucky heard a distant gobble. Otherwise, the morning was filled with the peaceful sounds of robins singing and an occasional chipmunk scurrying. Bucky dozed for minutes at a time, often laying his head flat on the ground. Around midmorning, the big buck became alert when he heard a twig snap in the woods. At first, he saw nothing. Soon, he saw that several deer were moving through the woods. Within minutes, they were out of sight, and the woods quieted again.

At midday, feeling hungry and rested, Bucky rose from

his bed and stretched. He had been bedded since before dawn, and it was time to head back to the clover field. As he moved through the woods, he encountered an area that was blanketed with trilliums. Immediately, he knew the clover field would have to wait. The white wildflowers were one of Bucky's favorite foods, and knowing they wouldn't be in bloom very long, he stopped to feast on them.

Silently, Bucky moved from flower to flower, nipping off the white beauties as he went. Periodically, he paused between bites to survey the forest for signs of danger. The only sounds were that of robins chirping and singing, and the occasional far-off gobble of a turkey. All was clear, so he continued to eat. Finally, feeling full, he walked through the thick hardwoods. When he got within 50 yards of the clover field, he bedded in a clump of deadfalls to wait for nightfall, for he knew the clover field would be safer after dark.

Though spring technically arrives on the vernal equinox in March, it arrives for whitetails when grasses and leaves begin growing. In Skipper's world, that isn't until mid- to

▼ Within the first hour of a fawn's life, the doe will feed and wash them, and then move the fawns away from the birthing site so predators cannot find them.

▼ When twin fawns are born, the birthing process takes about 45 minutes.

At birth, a fawn is about the size of a loaf of bread and weighs between five and nine pounds.

▲ Few natural foods are more preferred than apple blossoms in May.

◄ Wildflowers, such as these columbines, rank high on a whitetail's preferred food list when in bloom.

late May. The previous few days, Skipper had devoured the tender new grass shoots sprouting everywhere. Then, he received another treat, because leaves were emerging from buds on tree branches. Lush food was everywhere, and he took advantage of nature's bounty.

Dawn found Skipper high atop a ridge line overlooking a beautiful valley, where he was hastily feeding on the new leaves of a multiflora rose. As his mouth moved from leaf to leaf, he heard the distant gobbles of turkeys echoing from ridge to ridge. The early-morning sun felt good on his shedding coat. For the next hour, he nipped at the tender tasty leaves.

Just as Skipper thought of bedding, he saw two deer moving in thick brush below. Within seconds, he recognized it was Buttercup and Daisy, his mother and sister. He stared at them, not quite knowing what to do. Since April, when Buttercup had acted angrily toward him, he had wandered aimlessly, searching for a home. None suited him, so he returned to his birthplace. Cautiously he walked toward the does. He was hoping they would give him a warm greeting, but it was not to be.

Buttercup, heavy with fawns, pulled her ears back and strutted toward Skipper. This puzzled the small buck, because he couldn't understand why his mother was rejecting him. When Buttercup got within a few yards of Skipper, she started chasing him. In a flash, Skipper lunged sideways to avoid Buttercup's flailing hooves, and then bounded off through the hardwoods to avoid his mother's attack. After running a short distance, the young buck looked back to see Buttercup standing several yards away. Her body language told him that she was upset. Feeling rejected, Skipper began to walk and didn't stop until he was two ridges away. He couldn't comprehend what had occurred, but what he had experienced was Nature's way of ensuring that inbreeding doesn't occur in a whitetail herd.

Buttercup returned to Daisy, who was feeding on tender maple leaves. Daisy sensed her mother's tension but didn't know the cause. She was about to find out. With little warning, Buttercup turned on Daisy, like she had on Skipper. When Buttercup reached her daughter, she raised up on her hind legs, attacking her in a punching fashion with her front hooves. After being struck twice,

Daisy ran for safety. With Daisy out of sight, Buttercup returned to where she had been feeding. Quickly, the doe looked around for danger. Confident no predators were nearby, she walked into a thick tangle of brush she had scouted as a possible birthing site. She desperately needed to be left alone, because the fawns she had carried for the previous 200 days were about to enter the world.

Buttercup's contractions began before she could lie down. She quickly bedded. Uncomfortable, she stood and began licking her flank. Then, she emitted a guttural, barely audible moan. She was in the first stage of labor. As the painful contractions intensified, she squatted and arched her back as if to urinate. Then, her water broke, and the birthing process began. As Buttercup continued to lick her flank, a fawn began to emerge. Slowly, the fawn slid from the doe's body, landing softly on the ground. Immediately, Buttercup turned and licked the fawn to remove the amniotic sac from the little buck's face. A few strong strokes from his mother's tongue encouraged little Buttons to move.

By May, bucks begin forming bachelor groups and remain together until early autumn.

A buck's antler growth accelerates by the end of May, as food and sunlight increase with each day.

Vital Information

- In well-balanced deer herds, does begin birthing their fawns from late May to the first week of June, after a 200-day gestation period. In the weeks before fawning, does begin to aggressively defend preferred fawn rearing areas.
- In healthy deer herds, most does older than 2½ will produce twin fawns. About 53 percent of the fawns will be bucks.
- The average birth weight of fawns is about five pounds.
- After the birthing process and first feeding are complete, a doe will move her fawns to a safe location. This occurs during the first 90 minutes of their lives.
- Does will feed their fawns four to five times a day. During each feeding, a fawn consumes about 8 ounces of milk that's 10 percent fat.
- To keep predators from finding their fawns, does have them bed 40 to 70 yards apart.
- By the end of May, a whitetail's summer coat will have replaced its winter coat.
- If adequate forage is available, deer will feed extensively throughout the day, consuming 10 to 15 pounds of food.
- By mid-May, a mature buck's antler beams should be at least six inches long, with their brow tines visible.

Instead of tending to her needs, Buttercup washed the fawn with her tongue. As Buttons grew stronger, he began to move his legs, rolling his body from side to side. He wanted to get up but couldn't. With Buttons lying nearby, Buttercup again felt contractions. She was about to birth another fawn. This time, she was bedded when the little doe started to emerge from her body. To speed up the birth, Buttercup stood, causing Princess to free herself from Buttercup's womb. In 30 minutes, both fawns had been born.

With Princess freed from her amniotic sac, Buttercup began to lick her little doe, as she had Buttons. After both fawns were cleaned, the big doe tended to her needs by pulling the afterbirth from her body, chewing it and then swallowing all evidence of the birth. It was her way of ensuring predators would not find her or the fawns.

Within their first hour of life, Buttons and Princess attempted their first steps. Neither was very graceful, stumbling and falling as they attempted to reach Buttercup so they could nurse for the first time. Princess was the first to reach Buttercup. The little fawn fell to her knees, smelled for one of Buttercup's teats and started sucking. After nearly a minute of drinking her mother's rich milk, the fawn collapsed in a heap and curled up into a ball. Buttons was next, but he first had to get past his bedded sister, who was lying next to Buttercup. The little buck tried to step over Princess but didn't have the coordination to do so. He fell on top of her and rolled onto the ground. Quickly, he righted himself and crawled to his mother's side, nuzzled his nose between her legs and began nursing.

When Buttons had finished nursing, Buttercup rose from her bed and emitted a low nurturing grunt, which was her way of telling the fawns to get up. Neither wanted to move, but with a few soft grunts and a little nudging, Buttercup finally got them on their feet. While the fawns took turns nursing, Buttercup licked their genital areas to make them urinate — a step necessary to ensure their plumbing worked properly.

Less than two hours had passed since the fawns' birth, and Buttercup knew she had to move the fawns before predators found their location. The big doe walked away from the fawns and stopped several yards away. She turned to look back at the fawns while softly grunting at them, which was her way of telling the newborns to follow. As Buttercup walked away, Buttons and Princess hurried to catch up. Within the first 50 yards, each stumbled and fell several times. But the farther they walked, the more stable they became.

After walking the fawns several hundred yards, Buttercup reached a thick tangle of underbrush and stopped. She quickly surveyed the location before walking the fawns into the thicket, where she stopped and grunted softly. It was her way of telling Princess to bed down. After the little doe eased to the ground, she led Buttons away to find a safe place for him to bed. Within 50 yards, she found a safe spot for the little buck. She stopped, grunted softly and waited while he lay down. With the fawns bedded, Buttercup looked for a place where she could bed and hear the fawns if they needed her. After quickly scouting the area, she bedded a short distance from each fawn. It had been an exhausting day for Buttercup. It was time to regain the strength needed to nurse and raise her twins in the coming months.

CHAPTER 3

June

A thick blanket of fog floated in the valley
as the sun crept over the horizon. The night had been clear
and cool, which is common for early June. The first rays of sun
washed over Buttons' coat as they penetrated the thicket. Within minutes, the little buck felt a warming sensation that stopped
his shivering. He'd only been born a week earlier and was having
difficulty making it through the night without getting cold, so
the warm sunlight felt good. Two hours had passed since Buttercup had fed him, and he felt restless. With plenty of time,
he rose from his bed and peered into the green field nearby. He
could see several deer feeding. Curious, he walked toward them,
looking for his mother.

▲ For the first two weeks of a fawn's life, it remains bedded in a secluded location. The only time a fawn is active during this time is when the doe feeds it or moves it to another location.

▲ Fawns are very inquisitive and have no fear. This fawn was bedded on the edge of a clover field before it got up to check out a buck feeding nearby.

When Buttons got within 20 yards of the feeding deer, the closest yearling buck stopped eating and stared at him for a moment. The yearling had never seen a newborn and was curious about it. After a few seconds, the inquisitive buck cautiously approached Buttons. After reaching the fawn, the buck stretched out his neck and touched Buttons' nose, like a kiss. Buttons wasn't sure what to make of the encounter, so he just stared at the buck. The buck began to lick Buttons, much like Buttons' mother did when she fed him. At first, it felt good. But as the buck continued to lick Buttons, he also became more aggressive toward the fawn, gently pushing Buttons with his head. Unable to hold his ground, Buttons collapsed in a heap. The fawn struggled to rise, only to be pushed back to the ground by the buck.

Just as the yearling buck was about to cuff the fawn with his front leg, Buttercup appeared at the edge of the field and snorted loudly when she saw what was occurring. She had returned to check on Buttons after feeding Princess. Within seconds of snorting, Buttercup charged the buck. Before she could reach him, the yearling wheeled on his hind legs and bounded for cover, with Buttercup on his heels. With the buck out of the field, Buttercup gave up the chase and hurried back to Buttons. The fawn was a mess. His fur was matted and wet from rolling in the grass, and he was shaken from what had occurred. For the next few minutes, as Buttons nursed, Buttercup licked and washed the morning dew from the fawn's coat. After the feeding and grooming was complete, Buttercup walked her son back into the woods, where she forced him to bed down and stay put until she returned. It was time to see what Princess was doing.

Buttercup crossed a small stream and hustled up the bank. It had been more than four hours since she had fed Princess, and she knew the fawn would be hungry. Just as she crested the stream bank, she stopped in her tracks. Thirty yards away, near where she had left Princess, was a big male black bear. He was standing still with his nose in the air, sniffing something. It was fawning season but also the black bear breeding season, so it was not uncommon for boars to roam in search of a mate. Unfortunately for whitetails, that increased bear

The coincidence of fawning season with black bear breeding season results in many fawns being eaten by bears.

activity in late May and early June results in many new-born fawns being eaten by bears. Buttercup knew she always had to be on guard for bears when her fawns were born. And now, she had one in her lap.

Judging from how the bear was acting, Buttercup was sure the bruin had smelled the fawn. She knew if she didn't act quickly, he would probably find Princess and kill her. Buttercup snorted to get the bear's attention, hoping she might lure the bear into following her. The bear stopped and began walking toward Buttercup. The old doe knew she had to do anything necessary to lead the big boar away from Princess. When the bruin approached to within 15 yards, Buttercup turned to move away from it. If she faked an injury, the bear might think he could catch her. As she moved off through the woods, she dragged her right hind leg. The black bear took the bait and followed.

Each time the bear got within 10 yards of Buttercup, she moved just fast enough to lead it farther from Princess. After a few minutes of pretending she was injured, Buttercup had led the bear far from where they had started. Believing it was safe to run for her life, the big doe bounded off through the woods, making an arching circle away from the bear. She didn't stop running until she reached the edge of a large field, and then only for a moment. Buttercup knew she had to return to Princess and move her to another location if the bear decide to backtrack.

It only took Buttercup a few minutes to find Princess, who was still where her mother had left her. Time was of the essence. Buttercup softly grunted to make Princess stand. The fawn wanted to nurse, but Buttercup's soft grunting told the fawn that feeding would have to wait. Quickly, the doe and fawn moved toward the edge of the woods where a long hedgerow separated two large fields. On one side of the hedgerow was a clover field. On the other side, another field was being plowed by a farmer. Instinctively, Buttercup knew that the farmer's tractor

▼ During the first month of a fawn's life, a doe will feed and wash it four to six times a day.

▼ Food dictates where whitetails live during May and June. Usually, they bed within 500 yards of a prime food source.

▲ Whitetails are very nocturnal in June. Their daytime activity is mostly limited to the two hours on either side of darkness.

would keep the bear from venturing into the open, so the hedgerow would be an excellent place to hide Princess for the rest of the day.

The doe and fawn ambled down the back side of the hedgerow, out of sight of the farmer. After walking a short distance, Buttercup stopped to groom and nurse Princess before choosing a bedding site. Then the doe and fawn slowly walked into the narrow sliver of brush. When she found a suitable bedding place for the fawn, Buttercup emitted two soft grunts to tell Princess to lay down. Slowly, the fawn eased into a depression between two logs and curled up. For the next hour, Buttercup bedded in the hedgerow 50 yards away, between the woods and fawn, so she could protect Princess should the bear return.

Several times, the farmer's tractor came roaring by, turning the soil in preparation for planting season. The doe and fawn were hidden so well that the farmer never saw them. After the tractor completed another pass by the hedgerow, Buttercup got up and sneaked down the hedgerow to the woods. It had been more than three hours since she had checked on Buttons and rescued him from the yearling buck. She knew he must be hungry.

▲ During June, when antler growth is exploding and does are lactating, whitetails will consume up to 15 pounds of highly nutritious food per day.

Any preferred forage within reach of a deer is fair game. This photo shows how six-foot-high browse lines are formed.

Since daybreak, Bucky had been bedded in a thick apple orchard. Though he was within sight of a farm building, the big buck felt safe nestled in the thick cover. One reason he had chosen that bedding location was the fragrance from a stand of wildflowers growing in the orchard and tender pink apple blossoms he loved to eat off the branches. The location had everything he desired: food, cover and security.

Although it was quite warm for early June, Bucky didn't mind. His summer coat of thin reddish fur was nearly grown, so he was comfortable. As midday approached, Bucky rose from his bed, stretched and urinated. Before moving, he stood rock still, looking, listening and smelling in every direction to assess the danger. Convinced he was free from harm, he began to weave through the thick stand of phlox in the orchard. For the next hour, he filled his stomach with tender apple buds and blossoms and then bedded near the area where he had spent the morning. He intended to remain in the orchard until nightfall.

By late afternoon, the air became warm, and it started to get dark. In the distance, Bucky heard booming thunder, alerting him of an approaching storm. Wind began to blow, and within minutes,

By the end of June, most bucks are part of close-knit bachelor groups that bed, feed and move about their home range together.

large drops of rain started falling. As the wind and rain intensified, apple blossoms floated through the air like wind-driven snow. Soon, the raining blossoms covered Bucky's back. The orchard grew very dark, and lightning bolts flashed every few seconds, followed by thunderous booms. Bucky panicked. Soaked and covered with apple blossoms, he jumped from his bed and ran for the cover of the nearby forest. Within seconds, he reached a stand of giant hardwoods and quickly bedded, not sure what would happen next. Within minutes, the storm ended as fast as it began, and his fears subsided.

When fawns are four weeks old, they begin to accompany their mother at common feeding areas.

▲ By late June, mature bucks will begin to determine dominance within their bachelor group. Most of these encounters involve behaviors such as stare-downs, threat walking or all-out fighting with their front hooves.

▲ In prime farm country, up to 50 percent of a whitetail's June diet consists of natural forage such as the leaves on which these bucks are feeding.

It became much cooler, and as the sky brightened, a few shafts of sunlight broke through the clouds. The big buck had been through many storms, but this had been as intense a storm as he had ever experienced. With the wind gone, Bucky felt the urge to see what had occurred, so he rose from his bed and stretched. For a second, he paused before humping his back and shaking his body. Water jettisoned from his coat as he shook, resembling mist and fog being thrown from his body.

With his coat free of water, Bucky groomed himself for several minutes. Deer can lick almost every inch of their bodies, and Bucky didn't stop licking until every hair was back in place. He followed that with a quick massage of his growing antlers. Though he couldn't see his antlers very well, he knew they were starting to develop because he could feel them growing. The sensation prompted him to stimulate the antlers. The only way he could do so was to place one beam at a time under his hind leg and drag it to his hoof. Until

Fawns begin to show their independence by the end of June. They will often venture away from their bedding location without their mother.

Vital Information

• If food, cover and water are adequate, a doe will commonly stay in an area smaller than 100 acres during her fawn's first month of life.
• Fawns begin eating vegetation by the time they are 25 days old.
• During June, when fawns are lactating and bucks are growing antlers, they need more energy-producing food to remain healthy. Therefore, it's critical that foods containing 16 percent protein are available.
• DNA research has revealed that twin fawns can be sired by different bucks.
• Does locate bedded fawns by emitting a low-intensity maternal grunt that's audible for only a few yards. When a fawn hears its mother's grunt, it leaves its bed and runs to the doe's side to nurse.
• Fawns respond to their mother's calling by mewing. When in danger, a fawn alerts its mother with distress bleats.
• In prime farm country, a buck's summer range is often less than 600 acres.
• By mid-June, a mature buck's antlers will be heavy and bulbous. Its antler beams will be longer than 12 inches, and tines can be as long as six inches. From mid-June to mid-July, antlers can grow up to an inch a day.
• By the end of June, most bucks are part of close-knit bachelor groups that bed, feed and move about their home range together.

velvet peel in early September, that behavior would be common, occurring several times a day.

It had been several weeks since Skipper had seen his mother and sister, but he was doing well. Since being chased off by Buttercup, Skipper had located and become accepted by several other bucks his age. The loose-knit bachelor group consisted of four other yearling bucks, and the deer were usually close to each other. One of their favorite nighttime bedding areas was a fallow field that bordered an alfalfa field. It was dense with grasses and wildflowers high enough to conceal bedded deer. The bucks often bedded amongst the wildflowers, sometimes throughout the day.

Often, Skipper would groom other bucks in his bachelor group, and the other buck usually returned the favor. Such grooming sessions strengthened the bond this band had for each other. Little did they know the bond would be severely tested during autumn.

Daisy missed Buttercup. She had been separated from her for more than three weeks and felt an urge to look for her. As nightfall chased the last rays of daylight away, Daisy rose from her bed in a spruce plantation and headed toward a nearby alfalfa field. As she approached the field, she saw other deer already feeding in the knee-high forage. The farmer had not yet taken his first cutting, so the field was a great food source.

Daisy entered the field and began to feed. Sensing that something was entering the field behind her, she swiveled her head to look toward the sound. She immediately recognized the deer as her mother. Cautiously, Daisy turned around to face Buttercup. Their eyes locked, and the deer walked toward each other. It was as if they never had their scuffle the day Buttercup's fawns were born. Buttercup seemed anxious to renew their relationship. For almost 10 minutes, mother and daughter licked and groomed each other before feeding. They spent most of the night together. The only time they separated was when Buttercup left the field for an hour or so to feed Buttons and Princess.

CHAPTER 4

July

Buttercup's fawns were now big enough to eat solid food — such as flowers, clover and alfalfa — so it wasn't critical that they be nursed every few hours. Therefore, they didn't need their mother's constant protection. Since the end of June, Buttercup had allowed Buttons and Princess to follow her as she fed, bedded and avoided danger. As a result, Buttercup, Buttons, Princess and Daisy had become a close-knit family group by mid-July.

▲ Each time a doe nurses her fawn, the fawn will consume up to eight ounces of high-fat milk.

It had been hours since sunset. Fortunately for the deer, the temperature was dropping. The day had been extremely hot for July, and the deer could take a break from the constant harassment from insects. As Buttercup's family group fed in the big clover field, a full moon hung in the sky. The air was clear, making it easy for them to see the ocean of stars surrounding the moon. In the distance, a pack of coyotes squealed and howled, sending chills up every deer's spine. Although the coyote serenade was a nightly event, the deer never got used to their singing because of the threat coyotes posed to every creature.

The night was so calm that Buttons and Princess heard the "snip, snip, snip" of their mother biting off each stem of clover. Every so often, Buttercup's family group was alerted to events unfolding in the big field. The night's first act occurred when a family of raccoons got into a squabble. Between their hissing at each other, scurrying up and down a tree at the edge of the field, and constant fighting, they put on quite a show for the deer.

Just before midnight, Buttons and Princess witnessed something they had never seen before. In the shadows near the edge of the field, they saw small flickering lights that appeared to be moving. Buttercup and Daisy paid little attention. But the fawns, having never seen such a sight, moved in for a closer look. When they got to the edge of the woods, they were surrounded by lightning bugs, which turned their lights on and off as they hovered in the warm summer air. Buttons and Princess tried to catch the bugs, but every time they were about to do so, the bug's light went out, so the deer couldn't see them. After several minutes, the fawns lost interest and returned to Buttercup and Daisy to feed.

For more than an hour, Buttercup, Daisy and the fawns fed before bedding close to each other near the middle of the field. After a long, hot day, the does and fawns were able to relax. For the next few hours, they groomed and dozed as they waited for dawn. As they lay bedded in the field, Buttercup and her family heard the constant hoot of a great horned owl in the distance.

▼ Whitetail bucks are very protective of their antlers during the growing process, and by early July, a buck's antlers will be bulbous and half grown.

▼ Although the rut is more than four months away, a buck will test a doe's urine to check on her estrous status.

Just before dawn Buttercup, Daisy and the fawns were alerted by the distress bleats of a fawn. Although the bleating didn't seem close, they knew danger was lurking. When the bleating continued, Buttercup and Daisy jumped to their feet and came to full alert. The fawns followed. The deer stared toward the bleating. At the far edge of the field, a deer snorted. Buttercup was concerned. The bleating stopped, so Buttercup didn't know if the distressed fawn was dead or alive, but she wouldn't wait to find out. In unison, Buttercup and Daisy bolted for the safety of the woods, with the fawns close behind.

Dawn arrived windless, hot and humid — the kind of July day every white-tailed deer hates. Since first light, Bucky had been bedded in tall grass at the edge of the forest. At first, the density of the grass kept the insects at bay, but as the sun inched higher, tiny flying insects and horseflies started to size him up for a meal.

By midmorning the situation went from bad to worse. The hovering insects were everywhere. They were biting his face, ears and back, and some even tried to climb inside his eyelids. Bucky flicked his ears, jerked his head from side to side and twitched his coat as he tried to shake the intruders from his body. With

Water is crucial to whitetails' survival, and much of it comes from the plants they consume. Still, deer gravitate to any flowing water.

each minute, more flying creatures came to hover and bite. Unable to take it any longer, Bucky jumped from his bed and bolted through the tall grass. Even as he ran, he couldn't escape the fog of insects that followed. When he stopped in a clump of tall goldenrod, the insects pounced on him. He knew he had to find a cooler place if he expected any relief.

Running for his life, Bucky headed for the nearby forest, taking care not to snag his growing antlers on any branches. It was cooler and darker when he got a few yards inside the woods line. For a moment, he had left the army of insects behind, but he knew they'd find him if he didn't keep moving. Quickly, he walked down a steep embankment toward a stream. He knew it would be cooler there. When he reached the stream, he found a small waterfall that flowed into a pool. Without hesitation, he stepped into the knee-deep pool. For a few moments, he drank and then put his face into the water to sooth his stings. Next, he dropped to his knees and eased into the pool, letting the water flow over his body.

Bucks and does live separately except during the rut. About the only time they are seen together during non-rutting months is when they feed at prime food sources.

Bucky lay in the pool for several minutes before standing and shaking the water from his coat. After assessing the situation, he knew he should bed for the day, but where? A few feet from the little waterfall, he noticed a small gravel bar about the size of his body. After cuffing a few of the bigger flat rocks aside, Bucky bedded in the cool, moist pebbles. He was confident the location would keep him free of insects until the cool night arrived.

Soon after bedding, the big buck became mesmerized by the sound of water flowing over the small waterfall. Every so often, the water's gurgling was drowned out by the high-pitched whining of locusts as they worked in the treetops overhead. Though the sounds kept Bucky from going into a deep sleep, he dozed periodically for a few minutes at a time. By midafternoon, the day's intense heat finally reached the stream bed. Soon, a fog of small flies hovered around Bucky's head, with some landing on his nose and ears. Although he continually flapped his ears and shook his head, he couldn't get rid of them. Sleeping was out of the question. Bucky had to find another place to escape the onslaught of insects.

Insects take a heavy toll on whitetails during July. In some locations, the stress can be greater than what deer experience in winter.

A doe is very protective of her fawns and will seldom back down to a buck's aggressive behavior.

▲ Scent marking of a scrape's licking branch becomes more frequent as summer inches toward autumn. If the licking branch is in a prime travel corridor, all bucks in the area will leave their scent on it.

Bucky bolted from the stream bed, leaving the fog of insects in his wake. He didn't stop running until he reached a patch of thick goldenrod. He immediately bedded, hoping the flies wouldn't find him. Within a half-hour, his body was covered with flying creatures, and their stinging bites forced him to move again. Finally, nightfall arrived, and the temperature dropped, curtailing most insect activity. Bucky had survived another hot July day, but it had taken a great toll on his body.

⌒

Dusk was just settling in as the members of Skipper's bachelor group stepped out of the woods and poked down a long hedgerow. All were hungry after spending the day bedding and fighting off insects. They wanted to spend the night feeding in a big alfalfa field, but first they had to get there — hopefully undetected. Though it was less than a half-mile to the field, they needed to cross a pasture — which meant jumping two cattle fences — before crossing a gravel road.

The bucks effortlessly bounded over the fences as cattle watched them. As they approached the road, they heard a car coming. For a second, they stopped, not knowing what they should do. Though they had some protection

By the end of July, the antler growth of most mature bucks will be nearly complete.

from the nearby hedgerow, they didn't want to venture into the open pasture where the cows were grazing.

As if on command, the young bucks stayed close to the hedgerow and ran for the road. The deer and car were on a collision course. Because the brush concealed their approach, the driver didn't see them coming. Just as the bucks bounded into the road, the driver slammed on his brakes, causing the auto to skid on the gravel. Running at full speed, the bucks hadn't seen what had happened. However, the sound of the car was enough to keep them running across the field. Only after entering the woods did they stop. They were lucky to be alive.

The setting sun slid beneath the horizon as Skipper and his bachelor friends stepped into the alfalfa field. It had been a long, hot day, and they couldn't wait to fill their bellies. Soon after nightfall Skipper saw lights coming down the gravel road. He had seen vehicles pass by in the past while he fed at night, so he didn't feel threatened. When the car reached the field, it slowed, making Skipper look up to see what would happen. He heard sounds from the car about the same time a powerful beam of light shone toward him. A second later, the light moved to other parts of the field. Each time the beam settled on another deer in the field, Skipper heard sounds from the car's direction. He didn't realize

humans were driving the back roads, spotlighting for deer. After the car moved on, Skipper and the other deer spent the night in relative peace, listening to the drone of insects and far-off sounds of coyotes howling.

As dawn approached, Skipper spotted a lone deer entering the alfalfa field. He wasn't sure who it was, but from the animal's shape, he knew it was a big buck. As the buck fed closer, Skipper recognized it was Bucky. It had been a while since the two had seen each other, and Skipper gazed at Bucky's large antlers. The small basket antlers he and his bachelor group were growing looked pretty puny next to Bucky's.

Feeling inferior and intimidated by Bucky, Skipper didn't try to communicate with the big buck, although they fed within yards of each other. After a while, Bucky fed closer to Skipper, and then walked up to him and grunted softly. Sensing that Bucky wanted to be groomed, Skipper cocked his head sideways to avoid Bucky's huge antlers and cautiously began licking his front shoulder. Soon Bucky returned the favor, and for the next few minutes, the bucks groomed each other under the stars. As the gray of dawn began sweeping over the field, Skipper knew it was time for him and the other bucks to return to their daytime beds. Bucky and the young bucks parted ways, leaving the field in opposite directions.

CHAPTER 5

August

The night had been particularly dark, with cloud cover blocking out the stars. The air, heavy with moisture, alerted Bucky to a looming storm. After feeding heavily on clover for most of the night, Bucky headed for his favorite sanctuary to rest for a couple of hours before daybreak. His hiding place was a thick forested bench about 100 feet below a hardwood ridge. The area had many large beech trees and clusters of hemlock saplings, making it the perfect place for Bucky to hide. Though the thick cover limited visibility, the uneven terrain's thermals and constantly changing wind currents made it easy for Bucky to smell predators long before they detected him. Because of such bedding areas, Bucky had been eluding man and beast for more than six years.

▲ Grooming is a bonding behavior exhibited by all whitetails. It's taught by the doe during the first days of a fawn's life.

▲ August is a time of exploration for fawns. They spend a lot of time attempting to socialize with almost every deer they encounter.

When Bucky made it to the bedding site, the first hints of daylight were beginning to creep into the forest. He cuffed away a few sticks and leaves and then bedded. The heat and insect troubles of July were gone, so it was easier for Bucky to rest, though he knew there would be some days when it seemed like July had returned. Feeling safe and content, Bucky started chewing his cud. He closed his eyes and started to fall asleep, easing his head to the ground. His big velvet-clad antlers kept him from laying his head flat but supported his head enough to let him drift into a deep sleep.

As daybreak arrived, songbirds began to chirp and sing, waking Bucky. In the distance, he heard faint booms, signaling an approaching thunderstorm. As the booms grew louder, a pack of coyotes began to howl. Their serenade sent shivers down his spine. Seconds later, a flash of lightning lit up Bucky's bedroom, followed by a huge boom that resonated through the forest. The storm had arrived, and the wind began to blow. Though it was still relatively dark, Bucky saw the trees in his area swaying back and forth. A short distance away, a loud crack echoed above the howl of the wind. A big tree was falling, breaking branches from nearby trees as it fell. It hit the forest floor with a thud. Rain began to gush from the sky, moving sideways on the wind and soaking everything in its path. The noise of the wind and rain made it impossible to hear anything except the thunderous booms of the storm. For the next half-hour, lightning bolts flashed from the sky. Bucky had been through hundreds of such storms, but this one was as intense as he had ever experienced. Knowing there was no place safer than where he was, he chose to stay put.

It didn't take long for the thunderstorm to pass, and the forest calmed. Although the rain had ended, water droplets falling from the tree canopy made it sound and feel like it was still raining. In his bed, Bucky watched as thermals carried plumes of fog past him and up the ridge. Uncomfortable and soaked, the big buck stood and shook his body with all his might, propelling the water in every direction. Though he wasn't complete-

▼ If a mature doe feels stressed or threatened by a buck, she will not hesitate to show her dominance by fighting with her hoofs.

▼ A whitetail's world is driven by scent. Commonly thought of as a rutting behavior, bucks test a doe's urine throughout the year to determine breeding status.

Deer prefer few foods more than apples. High in sugar, apples remain a whitetail food staple as long as they are available.

ly dry, he felt much better. For the next few minutes Bucky stood licking and grooming his body. Every so often he paused to listen to the birds, for he knew their chatter would warn him if danger was nearby. With the storm ended, blue jays squawked back and forth to each other. On the tree branches around him, chickadees danced and chirped as they looked for breakfast. Sensing all was clear, the big buck moved a few feet and bedded again.

༄

The day was overcast and unseasonably cool for August — the kind of day that signals autumn isn't far behind. It suited Skipper and his bachelor group just fine. Midafternoon found them milling around an abandoned apple orchard. A few small golf-ball-sized apples had begun falling, so the bucks were scouring the ground to find an easy meal. Unfortunately, the search was difficult.

Weaving his way through the thick goldenrod, Skipper paused to test the wind. He didn't smell any apples on the ground, but he smelled them hanging on the trees overhead. There was just one problem: They were

In early August, forages such as clover and alfalfa are staples. Once apples and acorns begin dropping in late August, deer gravitate to these mast foods.

out of his reach. As he surveyed the situation, Skipper thought he might be able to pick an apple if he could get his mouth on one. Doing so would mean he'd have to stand on his hind legs, which he'd never done.

With opportunity knocking, Skipper decided to see if he could stand and balance himself long enough to pick the apple off the branch. Cautiously, he stretched out his neck and raised his front shoulders into the air. It didn't work. Determined, Skipper thrust himself upward. Though he reached the apple's level, he couldn't steady himself long enough to secure the fruit. He tried one more time. As he reached the apple's height, he took two small steps, smelled the apple, and latched onto it with his mouth about the same time his body began its descent. Halfway back to the ground, the apple snapped

free of the branch, making the branch and its cluster of apples whiplash into the air. When that happened, several apples broke away from the branch and fell to the ground. Hearing the apples hit the ground, the other bucks came to see what had happened and eagerly gobbled them up. The day was a blessing for Skipper. He got to eat one of his favorite foods and also learned how to find and pick apples when they weren't on the ground.

~

It had been a wonderful August day; clear and cool enough to keep pestering fly activity to a minimum. Sensing it was time to head to their favorite clover field, Buttercup and her family group rose from their beds and stretched. They had been bedded since midmorning, and with nightfall fast approaching were anxious to get something to eat. It was amazing how much Buttons and Princess had grown. In two-and-a-half months, they had gone from 5-pound newborns to whitetails nearly as big as her. It wouldn't be long before their winter coats grew in, erasing their spots.

With more than an hour of daylight remaining, the deer emerged from the woods and began feeding in the clover. A few minutes after beginning their feast, Princess

▼ The setting of this book is in an area like this, and the photo on page 166.

▲ After fawns are a month old, foxes pose little threat to them.

spotted the tall grass moving near the edge of the field, about 20 yards away. Within seconds, a gray fox popped out of the grass and walked parallel to her along the field's edge. Princess had seen foxes before but never that close. She didn't know what to think of the vixen. Curiosity finally got the best of Princess, and she slowly walked toward the fox, which was looking over its shoulder at her. When the fawn got within a few feet of the fox, it began walking in a circle around the fawn. Each time Princess tried to walk up to the fox, it jumped back. Their game of tag continued for several minutes before the fox had enough and slithered back into the high grass.

～

As Buttercup, Daisy and the twin fawns fed at one end of the clover field, three bucks entered the far end and began feeding toward them. With nightfall approaching, a lone deer entered the field. It was Bucky. Instead of eating, the big buck slowly approached the three other bucks feeding in the middle of the field. One of the three was almost the same size as Bucky and much more muscular than the two bucks he was with. His velvet-clad antlers were wide and tall, about the same size as Bucky's.

▲ When August is winding down, bucks will actively leave their scent on well-used licking branches throughout their territory.

If hot, dry conditions persist in August, deer seek refuge near water, where temperatures are cooler.

When Bucky got to within 20 yards of the three bucks, he bristled and pulled his ears back. He also changed the way he walked. He cocked his head to the side and strutted toward the bucks in a stiff-legged manner. The threat walk was Bucky's way of telling the bucks they weren't welcome and needed to leave. When Bucky had closed the distance to a few yards, he stopped, jerked his head upward and emitted a loud snort-wheeze sound. Bucky was telling them he would not tolerate them in his field, and they needed to leave — now.

Quickly, the two smaller bucks moved toward the woods. They had witnessed Bucky's aggression in the past and wanted no part of him. However, the bigger buck was new to the area and had no clue who Bucky was. It was clear that if the buck lingered much longer, he would soon find out what the old man was about. Just as Bucky was about to rush, the big buck backed down, wheeled and ran toward the other bucks, with Bucky in hot pursuit. After chasing the buck for 40 yards, Bucky slowed his advance to a walk, ushering the three bucks into the woods and out of sight.

▼ If a whitetail feels another deer is intruding on its food source, it will try to drive off the other deer.

▼ The quest for dominance increases dramatically toward the end of August. Even if a buck has not peeled its velvet, his antlers are hard enough for him to fight.

Vital Information

• In the North, a whitetail's winter coat begins replacing its summer fur by early August.
• Grooming is recognized as a bonding behavior and is carried out by bucks and does.
• By mid-August, many northern fawns weigh 60 pounds.
• It takes about six years for a buck's body to fully mature. In most cases, peak antler growth doesn't occur until a buck is six to eight years old.
• For most bucks, antler growth is complete by Aug. 10. For the next 20 to 25 days, the antlers go through a hardening process before the velvet peels.
• Although velvet still covers a buck's hardening antlers, aggressive antler-to-antler fights between bucks can occur in late August. Usually, antlers have hardened to a point they do not break.
• A small percentage of bucks will peel velvet in late August.
• From a food standpoint, August is a transition month for whitetails. In early August, forages such as clover and alfalfa are staples. But as soon as apples and acorns begin dropping in late August, deer gravitate to those mast foods.
• From dawn to midafternoon, deer prefer to bed on ridge tops, where they can take advantage of rising thermal currents brought on by increasing temperatures.
• From late afternoon into night, air currents begin flowing downhill. Deer bed in the lowest locations during this time.

Although he still had velvet-covered antlers, Bucky had proven he was willing and able to fight, even if it meant damaging his antlers before they were fully hardened and peeled. A change was in the air, and although it was still August, the big buck knew he had to secure dominance well before the rut if he expected to have breeding rights in November.

With the other bucks gone, Bucky walked along the field edge to one of his favorite overhanging branches. When he reached the branch, he arched his back, stretching his neck and head upward so he could run his nose and mouth along the branch's stub. After salivating on the branch, he rubbed it against his preorbital gland in the corner of the eye. He ended the ritual by splaying his legs and urinating into the ground. By making the scrape, Bucky let every deer in the area know of his presence — and dominance.

CHAPTER 6

September

As dawn broke, Bucky left the big clover field and headed for a nearby apple orchard. Apples had begun to fall from the trees, and the big buck wanted to eat a few before bedding for the morning. As he walked from tree to tree, Bucky's velvet-clad antlers brushed against thick honeysuckle bushes, splitting the velvet slightly from contact. He stopped, picked an apple off the ground and began eating it. As Bucky chewed, droplets of blood began oozing from his antler beam and dripped onto his ear. At first, it seemed like heavy dew from the leaves, but it was blood. Within seconds, Bucky became alarmed when the odor of blood drifted past his nose. He hated the smell, for it reminded him of danger.

Velvet peel is an exhausting process.
Though rare, bucks can die if they get their
antlers tangled in wire during the process.

Before he could swallow the apple, Bucky panicked. He wasn't sure what was on his antlers, but he knew he had to get it off. Frantically, he thrashed his rack against a thick honeysuckle bush. Immediately, blood trickled from the longest tine on his rack. Seconds later, he drove and twisted his antlers into the bush, trying to rid himself of the dripping blood. The foul smell of decaying velvet and dripping blood turned Bucky into a raging animal, and for the next 10 minutes, he took out his fury on every nearby bush.

The noise Bucky was creating prompted two other deer to approach and investigate. They stood and watched for a few moments before moving on. As Bucky thrashed his antlers against the brush, long strands of velvet came loose and hung from his antlers. Several times, he thrust himself into the brush, feverishly swinging his head from side to side. Each time that caused the longer strands of velvet to wrap around his antler beams or fall across his face. With each snap of his head, Bucky's face and back were streaked with more blood.

▼ Odors of blood and decaying velvet interest bucks.

▼ It's common for bucks to eat peeled velvet if they can locate it. The velvet contains high concentrations of protein, and it's thought that bucks exhibit this behavior to keep from attracting predators.

Exhausted from the ordeal, Bucky stood motionless for a few seconds. He still wasn't sure what was happening but knew he must rid himself of the bloody velvet. When his breathing returned to normal, he began to thrash the brush again. Frantically, he whipped his head and antlers clockwise against the brush, causing more tears in the velvet, which prompted more blood to flow. Bucky's heart was racing out of control, so he had to stop and catch his breath. When his panting slowed, he used his tongue to pull the hanging velvet strips into his mouth. Bucky began chewing the velvet to get the velvety skin free. It didn't work.

Scared from what was happening, Bucky bolted and ran through the orchard, with long strands of velvet streaming from his antlers. After running a short distance, he entered a ravine and headed downhill. When he reached the bottom, he stood for several minutes next to a small stream. He began to chew on the strands of velvet again. Bucky walked several yards to a cluster of hemlock saplings and began rubbing his

▼ After velvet is peeled, bucks begin to cover more territory. It's not uncommon for a buck's home range to encompass 2,000 acres in September.

▲ As October approaches and hormone levels increase each day, bucks begin paying closer attention to does.

head up and down on the soft-barked trees. When a long strip of velvet fell at his feet, Bucky stopped rubbing. For several seconds, he sniffed the velvet, and then picked it up in his mouth and began eating it. After a few chews, he swallowed the foot-long strip of bloody skin, a behavior most whitetail bucks exhibit during the velvet-peeling process to keep predators from finding them.

Exhausted from his velvet-peeling ordeal, Bucky collapsed in the hemlock thicket. It had been more than an hour since he had begun to rid his antlers of the bloody velvet, and he needed to rest. Unfortunately, the blood's odor kept him from bedding very long. Less than an hour after bedding, he was back on his feet, grinding away on the hemlock branches. By midday he had stripped his left antler of all its velvet. By late afternoon, almost all of his antlers were velvet free, with only one long strand of velvet still hanging from his right antler's base. The 10 hours of velvet shedding had taken a tremendous toll on the buck. Too exhausted to do anything but bed and rest, he waited to peel the final strand of velvet.

▲ During September, a whitetail buck has a dramatic increase in body weight if foods such as apples and acorns complement maturing farm crops.

By late September, mature bucks begin exhibiting dominance behaviors, like the stare-down this buck is imposing on another buck.

After velvet is peeled, bucks go out of their way to engage like-size bucks in sparring matches to determine dominance.

As long shadows crept over the big clover field, Bucky rose from his bed and stretched. Though he was sore and very tired from the day, he had to drink and eat. Slowly, he left the hemlock thicket and walked to the small stream. For the next minute, he drank from a clear, cool pool. When done, Bucky began walking to the clover field. He had worked up quite an appetite, and it was time to eat.

Skipper and his bachelor group had been feeding along the edge of the field for some time when Bucky emerged from the woods. The four young bucks stared at the big buck for a few seconds before continuing to eat. As Bucky fed closer, Skipper saw that Bucky had peeled his velvet, so he moved in for a closer look. When Skipper got within a few feet of Bucky, he stopped and tested the wind. He smelled the dried blood and strand of velvet on the big buck's antlers. Sensing Bucky's fatigue and irritation, however, he drifted off and returned to feeding.

Throughout the night, Skipper and the other bucks in his bachelor group stayed in the field. All had peeled their velvet and were feeling the effects of the hormonal changes in their bodies. Now and then, one of the yearling bucks walked to another member of the group and lowered his head, signaling that he wanted to spar. They would come together and spar for the next several minutes. Most encounters were civil, but at one point, Skipper and one of the other young bucks became embroiled in an ugly sparring match. For more than 10 minutes, they pushed and shoved each other back and forth in the clover. When Skipper caught an antler tine in his eye, he snapped. Viciously, he thrust his small basket rack into the other buck's side before locking antlers and continuing the tussle. Skipper quickly got the upper hand on his buddy, forcing the young buck to break off the fight and run for the woods.

▲ Sumac is a favorite food of birds and whitetails during autumn and winter.

With adrenaline flowing through his body, Skipper set his sights on one of the other bucks in his bachelor group; a 4-pointer that had been watching the scuffle from a distance. Skipper went after the 4-pointer with all the fury a buck exhibits during a November fight-to-the-death battle. Rather than engage Skipper, the 4-pointer turned and ran, taking with him the other small buck, which had watched from 30 yards away. Suddenly, Skipper was alone. For more than three months, the four young bucks had been inseparable. However, in a few minutes, Skipper had turned on his buddies. He didn't know it, but his rut had just begun.

⤴

Light rain fell throughout the day, dampening the countryside. It was also windy and unseasonably cool. To stay as dry as possible Buttercup, Daisy, Buttons and Princess spent the day bedded under a canopy of mature oak trees. Now and then, the wind picked up, making acorns fall and land on the forest floor near the deer. Several times during the day, Buttercup and her family rose from their beds, ate a few acorns and then returned to bedding. Food was everywhere. Clover and alfalfa forage was lush, apples and acorns were falling from the trees in droves, and wild grapes hung heavy from vines. It was a great time to be a white-tailed deer. With food in abundance, deer gorged themselves on nature's bounty.

The day's cool, heavy overcast conditions were comforting to Buttercup and her family. The respite from the recent warm September days made the deer head to the clover field early. They arrived more than two hours ahead of nightfall and quickly began feasting on the lush clover. Within minutes, other deer filtered into the field, and before nightfall, more than 20 deer were feeding there.

As darkness fell, Buttons and Princess interacted playfully with two other fawns in the field. It didn't take them long to become immersed in a deer's version of catch me if you can. One fawn and then another would bolt and run helter-skelter around the field, weaving between the feeding deer. Soon, two does joined the chase. The deer ran as fast as they could, almost as if trying to top each other. After 10 minutes of running back and forth, exhaustion set in, and the deer returned to feeding.

A mature doe can be very aggressive toward an immature buck during autumn.

Researchers believe early-season sparring is essential for determining hierarchy in the buck population. Sparring begins when a buck presents his antlers to another buck. Then they come together and push each other around in a nonaggressive manner.

▲ Running full speed, a buck can easily leap more than 20 feet when jumping over an obstacle such as this cattle fence.

Two red squirrels chattered excitedly as Bucky ambled down the narrow farm road that connected a clover and corn field. A light frost clung to the grass in the roadway, signaling autumn's arrival. As the first rays of sunlight painted the treetops with a soft brush of red, Bucky stopped along the edge of the road. For the next minute, he rubbed his nasal, preorbital and forehead glands on a maple branch overhanging the road, and then splayed his legs apart before urinating into the ground.

As Bucky was about to enter the clover field, he saw movement to his left. Another deer, a 2½-year-old 8-pointer, was walking parallel to him in a goldenrod field. The 8-pointer slowed, inquisitively watching Bucky as he neared him. Bucky quickly recognized the buck from past encounters. After scanning the clover field for danger, Bucky changed course and headed toward the 8-pointer. Cautiously, the smaller buck remained still as Bucky approached. He decided that if the big buck turned aggressive, he'd bolt. If not, he'd stick around to see what might unfold.

With the early-morning sunlight bathing the gold-

Vital Information

- For most Northern bucks, velvet peel occurs from Sept. 1 through 10.
- Most bucks peel their velvet in less than 12 hours. In some cases, the process can take days.
- After velvet peel, scrape behavior increases dramatically. The scrape-making process has three behavioral patterns: scent marking the overhanging branch, pawing the ground beneath the branch and urinating into the pawed ground.
- Most scrapes are in open, level hardwoods, along well-used trails and field edges.
- Bucks and does work the same scrape if it's in an area frequented by many deer.
- Dominance scrapes don't have an overhanging branch and are made when a buck aggressively paws the ground to show his dominance to another buck.
- Most rubs on trees more than four inches in diameter are done by bucks older than 2½.
- A greater concentration of rubs will occur in thick cover with small saplings.
- Researchers believe early-season sparring is essential for determining hierarchy in the buck population. Sparring begins when a buck presents his antlers to another buck. Then they come together and push each other around in a nonaggressive manner.
- During September, bucks and does can consume up to 15 pounds of food per day. It's not uncommon for a buck to increase his body weight up to 25 percent by late October.
- Fawns are the last to grow their winter coats. By mid- to late September, their spots will have disappeared.
- Daytime temperatures dictate the amount of daytime activity deer exhibit during September. Most of their activity occurs during the last two hours of daylight.

Here, a buck smells a scrape's overhanging licking branch to identify other bucks that might have left their scent.

enrod field, Bucky walked to the 8-pointer and stared into his eyes at close range. This let Bucky test the other buck to see if he respected Bucky's dominance. The 8-pointer calmly flicked his ears and slowly bobbed his head up and down, as if to tell Bucky he knew who the boss was. With Bucky continuing his stare, the smaller buck turned his head at a 90-degree angle, slid his rack under Bucky's massive antlers, and began to groom the big buck's neck and shoulders. After almost a minute of being massaged, Bucky returned the favor, and for next several minutes, the bucks groomed each other in the early-morning sunlight.

Their grooming was interrupted by the cutting of turkeys scolding each other in the nearby woodlot. Af-

ter the cutting turned to intermittent yelps, the bucks calmed and looked at each other closely. The 8-pointer backed up two steps and lowered his head, signaling to Bucky that he'd like to spar. The big buck obliged, and for the next few minutes, the click and clack sound from antlers coming together were audible throughout the field. Their sparring match ended when the bucks were alerted to the sound of an approaching tractor. Just as the tractor came into view, Bucky and the 8-pointer bolted. Each bounded in different directions as they ran for cover. Though they had almost daily encounters with farm equipment, they knew the machines were their predators and wanted no part of them.

CHAPTER 7

October

As the dark night eased to the gray of dawn, Buttercup groomed herself in a bed of freshly fallen maple leaves. Twenty yards away, Daisy, Buttons and Princess were curled up sleeping. It had been a month since Buttons and Princess had shed their spots, and they were nearly as large as Daisy and Buttercup.

As the old doe listened to the cawing of crows overhead, she thought about the encounters with bucks she knew would occur during the next two months. Her previous seven autumns had taught her that October and November were exhausting and dangerous.

Most rubs made during early October are done by mature bucks.

It had been a beautiful October day; calm, clear skies and cool for that time of year. As the last hint of red faded from the sky, Buttercup, Daisy, Buttons and Princess began feeding in their favorite clover field. After a few minutes, the old doe paused to look at Buttons and Daisy, who were grazing nearby. Buttons looked up, noticed his mother's gaze, rushed to Buttercup's side and began nursing. Feeling left out, Princess stopped eating and joined Buttons. The fawns nursed briefly and then returned to foraging on clover. Based on the way the evening had started, the four-some sensed their night would be stress free.

Soon after nightfall, two yearling bucks entered the field and began feeding nearby. Minutes later, one of the bucks, a small 4-pointer, walked to where Buttercup had urinated a half-hour earlier and sniffed the ground. After a few seconds, the young buck raised his nose into the evening air and lip-curled to determine if the urine was from a doe in estrus. When he finished testing the urine's odor, he studied Buttercup for a few seconds before approaching her. When the young buck was within a few yards, he broke into a bird-dog-trot; a half stalk, half trot whitetail bucks often use to approach does before the breeding season.

▼ After wild grapes ripen, they become one of the whitetail's preferred foods.

▼ Except during eating sessions, bucks and does are segregated in early October.

▲ Abundant food during September and October lets white-tails easily consume up to 15 pounds of food per day.

▲ Rubbing behavior intensifies as autumn progresses. Rubs are a way for bucks to leave their scent and visual calling card, which alerts other deer to their presence.

What happened next startled the 4-pointer. He thought Buttercup would run, allowing him to chase her. She didn't. The old doe had been part of that drill numerous times, and she wouldn't let a yearling buck dictate the rules of engagement. When the 4-pointer saw that the doe wouldn't run from him, he came to a screeching halt within feet of Buttercup. As if on cue, the old doe stood on her hind legs and flailed her front legs like a boxer. Before the little crotch-horn buck turned to run, Buttercup landed several blows to his neck and shoulders. Stunned and humbled by the experience, the buck wheeled and ran for cover, taking the other buck with him.

Had the pursuing buck been mature, Buttercup would have run when he approached. But she knew that if she ran from the young buck, it would open the door for other young bucks to chase her when she encountered them, and she didn't want that to happen.

Bucky's day dawned crisp and clear. Throughout the night, he had fed and bedded in a vast corn field. When daylight arrived, it was time to seek thicker cover. He picked through the outside rows of corn and walked into the woods. Immediately, he stopped dead in his tracks. Something was wrong. He could smell danger but didn't know where it was. Slowly, he looked around and moved his nose from side to side to test the thermals wafting back and forth in the woods. Instantly, his eyes saw something odd hanging on the side of a big oak tree a few yards away, about 20 feet above him. Seconds later, he saw a slight movement and heard a soft screeching sound, like that of metal rubbing against metal. Immediately, his mind matched the smell with the sight. It was man. In one motion, the huge buck turned on his hind legs, bolted back to the edge of the cornfield and ran for his life. After covering 100 yards, he darted into the woods and stopped. Though the smell of man was gone, he knew he must quickly distance himself from the area to ensure he didn't encounter any more of his chief predators. He headed for a thick apple orchard he knew would give him cover for the day.

▼ By the end of October, a whitetail buck's home range might cover up to 4,000 acres.

▼ In acorn-rich areas, whitetails will shift their home range to take advantage of this carbohydrate-rich food source.

By the end of October, scraping behavior is near its peak. When working a scrape's overhanging branch, a buck deposits scent from his nasal gland, preorbital and forehead glands on the branch. As with rubbing, this behavior telegraphs the buck's presence to other deer.

Bucky descended into the ravine, crossed a small stream at the bottom and slowly walked uphill. With each step, his nose processed the air for danger. All was clear. At the top of the ridge, the big buck walked on an old logging road that led to the orchard. Before reaching his destination, Bucky stopped to smell a cherry branch hanging over the road. The spot was a favorite scraping site, and as he ran his nose up and down the branch, he detected the odors of several deer. After he identified which deer had worked the branch before him, he rubbed his nose and facial glands on the branch. He heavily lathered the branch with his scent, and then added his exclamation point by pawing all the leaves away and urinating on the ground beneath the branch.

Before continuing, Bucky eyed a hemlock sapling a few feet away. He walked to it and shaved away a strip of the tree's tender bark with his incisor teeth. When the fresh aroma of the tree's core wafted into his nose, he began rubbing. For the next 10 minutes, the huge buck rubbed with all his might. Periodically, he stopped to lick the rub and then continued rubbing. When he was

▶ During autumn, does feed heavily and stay within their family groups. A doe's home range during autumn is typically 600 acres or less.

By mid-October, bachelor groups that formed in spring and summer will break up.

done, several of the tree's low-growing branches were broken off, leaving a two-foot-long scar on the side of the tree. With shreds of hemlock bark hanging from his antlers, Bucky continued his walk to the orchard.

The day was less than two hours old when Bucky reached the orchard. He veered off the roadway into a sea of goldenrod that blanketed the area. The goldenrod was taller than he was, making it tough to see very far ahead. Despite that, he smelled the apples and other deer. It didn't take him long to find both. The ground beneath the first tree was covered with apples. A doe and fawn were already feeding on the fallen fruit and stopped chewing to study the heavy-racked buck. With his eyes locked on the deer, Bucky emitted a low, guttural grunt before taking several quick steps toward the doe. Wanting no part of the buck, the doe and fawn quickly scampered off through the thick tangle of brush. More interested in food than does, Bucky stopped his pursuit and returned to the tree to feed. For the next 20 minutes, he consumed one apple after another before finally feeling full. It was time to bed.

For several hours, Bucky lay in the goldenrod, sheltered from the warm October sun by the canopy of the surrounding apple trees. Throughout his rest, the only audible sounds were the grinding of his teeth as he chewed his cud and the intermittent sound of crows cawing to each other. Unexpectedly, he was alerted by, "Jeer, jeer, jeer, jeer!" Squawking blue jays were fussing about something, but what? Bucky turned his head and looked toward the sound. When the jeering continued, he stood, trying to see through the thick goldenrod. Seconds later, the stillness of the calm October day was punctuated by a snorting deer. Something was wrong, and the big buck didn't like what he was hearing. A branch broke, followed by the sound of running deer coming his way. Bucky could barely see two deer as they ran past in the tall goldenrod. Just as he was about to bolt, a domestic dog came into view. It was following the deer that had just passed Bucky, and the dog never saw him as it ran through the orchard. The big buck had experienced many encounters with domestic dogs, and he knew the danger they posed. As soon as the dog was out of sight, Bucky bounded for safety in the opposite direction.

As autumn inches through October and bucks compete for the alpha position, sparring and fighting become intense.

During autumn, bucks are continually checking scrapes and rubs to determine which bucks are using their core area. This buck is smelling a rub made by a mature buck.

By the end of October, it's common for sexually active bucks to cover more than 3,000 acres as they search for does.

Skipper's hormone level was so high that he couldn't rest for very long. His obsession with does had replaced his desire for food. As nightfall settled, he rose from his bed at the edge of a swamp and walked aimlessly. He encountered many deer on his journey, and in the middle of the night, he and a basket-racked 6-pointer sparred aggressively. It began as a friendly sparring match but quickly became ugly. The sound of the bucks fighting attracted the interest of other bucks, including a mature 8-pointer, who tried to join the fight. The yearlings were so focused on each other that they didn't notice the big buck's approach. Without warning, the 8-pointer slammed his rack into the side of the yearlings' locked antlers. For several seconds, the three bucks' antlers locked, and their bodies rotated in a clockwise fashion.

▲ When scraping behavior peaks, sexually active bucks might make up to 10 scrapes every hour they are on the move.

When the yearlings realized what was occurring, they quickly separated and broke off the fight. They bolted and ran several yards before stopping to look at the big 8-pointer, who was still standing where the skirmish had happened. The big buck let out a long aggressive wheeze, and then walked toward Skipper with his ears back and head cocked sideways. Skipper bounded for the woods, waving his long tail from side to side as he ran.

As the night progressed, Skipper stopped several times to feed on acorns, apples and corn. None of the feeding sessions lasted more than a minute or two, as the young buck was more consumed with locating does than eating. Every time Skipper found a doe, he tried to see if she would accept his advances by lowering his head and trotting toward her. Each time, the doe would run off or become aggressive toward him.

With dawn breaking, the young buck walked into a wide hedgerow that separated two corn fields and bedded next to a fallen oak tree. As the sun inched above the horizon, Skipper heard leaves rustle on the other side of the log. He twisted his head to see what it was. A gray squirrel appeared atop the log and sat

▼ If daytime temperatures are higher than normal, deer will bed extensively. The warmer it is, the less active deer will be.

Vital Information
• Research has shown that whitetails actively feed when the barometric pressure is between 29.80 and 30.29.
• Scraping is recognized as instinctive behavior and increases dramatically with age.
• The most active daytime scrape sites are along trails in cover that connects bedding and feeding areas.
• During October and November, sexually active mature bucks can make up to 10 scrapes every hour they are on their feet.
• From mid-October through November, rutting bucks can make 10 or more rubs per day.
• If yearling bucks dominate a buck population, the seeking, chasing and bonding behavior associated with breeding will be very unpredictable.
• By the end of October, it's common for sexually active bucks to cover more than 3,000 acres as they search for does.
• Aggressive fights are common when there's a high percentage of bucks 3½ years or older in a well-balanced herd. In this type of herd, bucks respond well to antler rattling and calling.
• By the end of October, buck fawns in the North can weigh 70 to 90 pounds.
• When the air temperature increases above the seasonal norm for an area, deer activity decreases dramatically. The warmer it is, the less deer move.

on its haunches, mere feet from Skipper's head. Unconcerned with Skipper's presence, the squirrel held a corn cob in its paws and began to break kernels free. For the next few minutes, the squirrel ignored Skipper as it consumed kernel after kernel. When the last kernel of corn was gone, the squirrel dropped the cob and watched it roll off the log. Only then did the squirrel take interest in Skipper. For several seconds, the squirrel stared statuesque at the bedded buck, with only its long, bushy tail darting from side to side. Then, as fast as it had come to the log, the squirrel jumped off and was gone.

An hour after daybreak, the sky darkened, and the wind began to blow. Although the fallen log served as a wind break, Skipper felt the temperature dropping. A storm was approaching, and as the minutes passed, the wind became more intense. Nearby cornstalks rattled as leaves were blowing everywhere. Some leaves were helicoptering to the ground, some were going up, and some were blowing sideways. It was raining leaves in every direction. Big drops of rain began pelting the bedded buck, followed by wind-driven sheets of rain. Beads of water streamed from Skipper's thick winter coat and cascaded to the ground. Twenty minutes later, the rain subsided, and the wind calmed. The storm departed as quickly as it had arrived. Skipper rose from his bed, stretched and shook the rain water from his coat. He had been bedded for nearly two hours, and it was time to look for does.

CHAPTER 8

November

"Whooo." The low-gliding owl jolted Bucky from his sleep. Half awake, he watched as the barred owl landed in a nearby beech tree. For several minutes, the big buck trained his eyes on the moonlit raptor as it surveyed the forest. With the full moon casting long shadows, the owl sang, "Who cooks for you, who cooks for you." Each time the owl trumpeted, a chorus of hoots echoed down the ridge.

For almost a week of very cold nights, November's rutting moon shone brightly, prompting the bucks in Bucky's home range to be very active. The rut was about to explode. With dawn fast approaching, Bucky rose from his bed, stretched and urinated. For several seconds, he stood motionless, sifting the air for smells of danger. The only sign of life in the forest was the owl's eerie serenade. The big buck had been resting for more than two hours, and the urge to locate an estrous doe dominated his thoughts.

Bucky's antlers glistened in the light of the full moon as he began to walk. Every few yards, the big buck stopped to smell the air and listen for sounds in the fading night. Although he didn't expect to find any does until he reached the big cornfield in the valley below, he knew from experience that anything was possible. A quarter-mile into his trek, he left the ridge top and descended into the river bottom. The steep incline of the hillside made walking difficult, and the heavy blanket of leaves rustled underfoot. Bucky reached the bottom of the hill and crossed a hardtop road. The only audible sounds were the soft tap-tap-tap of Bucky's hooves landing on the pavement and the flow of the nearby river that paralleled the highway.

It took only a few seconds for Bucky to cross the road and step into the cold river. As he scampered through the knee-deep current, water splashed on his body. After he reached the other side, Bucky paused momentarily to shake water from his coat before catapulting his 250-pound frame up the well-worn deer trail on

▼ Fearing harassment from bucks, does are very skittish near food sources in November.

▼ When does near estrus, they will accept a buck's presence.

During the peak of the rut, bucks actively pursue all does in their area to find one to breed. By trapping the scent from a doe's urine in his nose and flehmening, a buck can tell which doe is in or coming into estrus.

the river bank. Stones clattered as he dug his hooves into the trail. At the top of the bank, the big 10-pointer paused to survey the sea of corn before him. It was here he hoped to find a doe to breed.

Bucky couldn't see any movement in the cornfield, so he listened intently. He knew that if any deer were feeding, he'd hear them. Immediately, he picked up the familiar sound of rattling cornstalks. Deer were feeding near the middle of the field. The big buck cautiously walked along the end of the rows before entering one he thought would lead him to the deer. Though he tried to be quiet, the narrow rows made it impossible to walk silently. As Bucky neared the deer, the sound of corn leaves rattling off his antlers spooked them. At the sound of the first cornstalk being flattened, Bucky paused and listened closely, hoping to determine where the fleeing deer were headed.

◄ By November, standing corn has matured, making it one of the whitetail's preferred foods.

▼ During the 48 hours before a doe comes into estrus, bucks will aggressively pursue her.

When the sound began to fade, Bucky walked to where the deer had been. Knowing urine would reveal any doe's estrous condition, he milled around for a few minutes, smelling the ground for any sign of urine. At several locations, he drew urine odor into his nose and lip-curled. Unable to detect any does in estrus, Bucky quickly moved through the cornfield. He had a lot of ground to cover before it was time to bed.

Exiting the field, Bucky walked to the edge of the woods and stopped under a stub of an oak branch that hung above his head. He reached up to touch it, arching his back to gain more height. Feverishly, the monarch rubbed his nasal, preorbital and forehead glands on the overhanging branch. The sound of his massive antlers whacking against it was audible everywhere. After working the branch, Bucky pawed the ground beneath it several times before slowly positioning his hind feet under his body so his tarsal glands touched. As the buck rocked back and forth, urine rained down over the rank tufts of tarsal hair, saturating the ground. The process of working the branch, pawing the earth and urinating had only taken moments, but the odor left behind would alert every deer in the area that he'd been there.

Although they do so infrequently, some does leave their scent on buck rubs.

With the scent-checking process complete, Bucky slowly entered the dark woods and began following the hot trail left by the does he had spooked in the cornfield. Twice in the first 100 yards, he stopped to examine urine they had deposited in the leaves. Convinced the does were not coming into estrus, the big buck lost interest and began zig-zagging through the woods, hoping to find scent from other does.

With dawn's gray shadows giving shape to the forest, Bucky stopped to feed on acorns. It had been hours since he had eaten, and it felt good to feed again. He crunched down on one acorn after another, as their hulls fell from the side of his mouth to the forest floor.

Just as he was about to reach for another acorn, he glimpsed a deer heading toward him. Bucky knew by its gait the buck was a small 10-pointer he knew well. They had spent many nights together during summer feeding in a clover field, often grooming each other. But now that the rut was heating up, things were different. With competition building for breeding rights, their days of being buddies were finished.

A dominant buck smells a rub to identify any buck that might have left his scent.

When the subordinate 10-pointer walked within 15 yards of Bucky, the big buck dropped his ears, brought his head up and approached the lesser deer. Hair on the massive buck's neck and back bristled. Then without warning, Bucky let out a loud, drawn-out wheeze. Instantly, the small 10-pointer wheeled and ran through the woods, breaking branches and scattering leaves as he bounded off.

Bucky watched intently as the bounding buck's white tail vanished. The big buck listened to the incessant cawing of crows in the distance as they greeted a new day. He also heard the faint sounds of turkeys as they yelped back and forth on their roosts. The new day would bring lots of activity from gray squirrels scampering about in search of acorns, so the big buck decided to get some rest before the noise of the day began.

Bucky was alerted to leaves rustling just as he dozed off. The buck spotted movement in front of him. Several does and fawns were moving through the woods, feeding on acorns. Overwhelmed by the urge to breed, the big buck sprang to his feet and stood motionless, surveying the situation before moving toward the does and fawns. With a blue jay scolding in the distance, Bucky approached the deer. As he stared at the does, Bucky picked up movement behind them. A buck was walking toward them, silhouetted against the sun on the horizon. Bucky recognized him as a mature 8-pointer he had encountered many times.

▲ This buck leaves scent on the overhanging licking branch before rub-urinating in the scrape below the branch.

The does and fawns trotted through the woods as the bucks approached each other. Bucky took the initiative by lowering his head, dropping his ears and bristling his coat like a pin cushion. The 8-pointer slowly eased around Bucky before dropping his ears and bristling his fur. Each buck rolled his eyes and cocked his head. The rut's fury was about to explode in the oak flat.

With the does and fawns watching from a distance, the bucks came together with all the energy their bodies could muster. The sounds of breaking branches, rustling leaves and clashing of antler tines was deafening. With their antlers locked, Bucky drove the 8-pointer into a small hemlock sapling, making the smaller buck lose his balance. Furiously, Bucky bulldozed the stumbling buck backward for several yards across the forest floor. With the 8-pointer's side exposed, Bucky rammed his antlers into the smaller buck's flank. Hair flew as one of his antler tines penetrated the 8-pointer's flesh, drawing blood. With adrenalin flowing full throttle, the 8-pointer struggled to his feet just as Bucky was about to move in for the kill. Despite his injury, the 8-pointer bolted through the woods with Bucky on his tail. Within seconds, the big buck broke off the chase.

Too exhausted to move, Bucky stood motionless, panting in the early-morning sunlight. Each breath resembled a blast of steam being shot into the air. After a few moments, the does and fawns wandered off through the woods. Just as they faded from sight, three subordinate bucks that had heard the fight approached and began milling around. Bucky looked toward them and uttered a loud snort-wheeze. The aggressive vocalization was enough to make the smaller bucks disperse.

With the does still on his mind, Bucky followed their trail, which led him to the edge of an abandoned apple orchard. The fight had taken its toll on the big buck, and it was time to bed and let his body recover.

Because acorn-producing areas are prime doe feeding locations, bucks will cruise these areas looking for estrous does.

During heavy rain, whitetails periodically shake their bodies to rid themselves of water on their coat.

The intensifying rut made each day increasingly chaotic for Buttercup, Daisy and the fawns. They couldn't make a move without a buck chasing them. If they ventured to their favorite corn or clover field to feed, they were chased. Even when they tried to bed in a dense cedar swamp near the river, bucks would force them from their beds. Rather than move, they hid in the thickest cover they could find to steer clear of the bucks' rutting wrath.

The day had dawned overcast and cold. About midmorning, wind began blowing though the spruce plantation in which Buttercup, Daisy, Buttons and Princess had taken refuge. The dense vegetation might provide cover from the elements and bucks, but that was not to be.

About midday, Bucky rose from his bed, stretched and urinated. Quickly, he tested the wind and then began walking toward the spruce plantation where Buttercup, Daisy and the fawns were bedded. It didn't take Bucky long to cross the field. Just as he entered the spruces, the wind shifted, and he picked up the scent of Buttercup and her family. Although he could not see them, he knew they were nearby. Slowly, the big buck picked his way through the dense growth of spruces, following the lead of his nose. Buttercup was the first to see the approaching buck, and she knew she was in trouble. When Bucky got close, the old doe sprang to her feet and ran, taking the other deer with her.

Bucky paused before taking up the chase. Within seconds, he plowed through spruce boughs and honeysuckle bushes, trying to get a glimpse of the deer he had jumped. Not wanting to leave the spruce plantation, Buttercup and Daisy stopped in a cluster of blown-down trees. Not knowing what to do, Buttons and Princess stood to the side and watched as Bucky appeared. Slowly, Bucky walked to the edge of the log pile, stretched out his head and looked into the eyes of Buttercup and Daisy.

For the next minute, the buck and does just looked at each other. The entire time, Bucky emitted a continuous flow of medium-volume grunts, known to biologists as tending grunts. Periodically, he stomped his front hoof on the ground. Drool was dripping from the corner of his mouth. He was ready to breed, but Buttercup and Daisy were not. With the big buck standing guard, Buttons and Princess tried to walk to Buttercup. Bucky took two steps toward the fawns, uttering a long, loud wheeze and then bluff-charging Buttons and Princess. His aggressiveness sent them running in opposite directions.

With Bucky riveted on the fawns, Buttercup and Daisy sprang from their beds and ran for their lives. The massive buck tore around the blowdown in hot pursuit. As the does and buck ran from the spruce, the sound of branches breaking was impressive. After running a short distance through open hardwoods, the does came to a narrow stream. Buttercup cleared the water in one bound, but Daisy missed her stride, and splashed water and clattered rocks as she crossed in front of the leaping Bucky. Buttercup immediately found shelter in another blowdown; a big cherry treetop with a tight crevice only big enough for her. Leaves flew as Bucky came to a screeching halt next to the blowdown Buttercup was in. For the next few minutes, the deer stared at each other. Daisy was nowhere to be seen, nor were the fawns.

A buck will stay with a doe for up to 72 hours during her estrus cycle: 12 to 24 hours before she enters estrus, 24 hours when she's in estrus and 12 to 24 hours after she cycles out of estrus.

The noise of the chase had attracted other bucks, and competition was quickly forming. With Bucky standing less than 10 yards from Buttercup, the clatter of rocks in the nearby stream made Bucky turn to see what was coming. Skipper and a 2½-year-old 8-pointer were cautiously crossing the small stream. Bucky snapped. He wheeled and ran toward the bucks, grunting with each step. The 8-pointer nearest to Bucky took the full brunt of the big buck's wrath. As Skipper and the 8-pointer wheeled and tried to run away, Bucky rammed his antlers into 8-pointer's flank, causing hair to fly. The buck stumbled but was quick enough to stay just ahead of Bucky as the big buck chased Skipper and him through the woods. Bucky chased them for a long distance before giving up. With the bucks momentarily out of sight, Buttercup left the blowdown and headed through the woods. Bucky soon returned but decided to not pursue Buttercup, as she was not in estrus.

Dominant bucks bed very little during the peak of the rut. When they do, they seldom stay bedded longer than three to four hours unless they are with an estrous doe.

When a doe is in estrus, a buck will attempt to make her stand so he can breed her.

After standing next to a huge oak tree for several moments, Bucky walked west through the woods. It had been some time since the big buck had left his bed. Although he was sleep-deprived, he remained energized because of a healthy dose of adrenaline. Every 100 yards or so, he stopped to work a scrape. Three times, he paused long enough to shred small saplings before moving on. Several times during his morning travels, he encountered does, and twice, he chased them a short distance before ending his pursuit. He was on a mission to find and breed an estrous doe.

Almost 10 days had passed since the November full moon had been full. With the smell of does floating on the wind, Bucky was ready to greet a new day. A cold, steady breeze blew from the northwest. As a mixture of rain and snow fell, Bucky emerged from the standing cornfield. This was the big buck's shining moment. The rut was about to heat up.

As Bucky entered a woods that descended into a deep ravine, he paused to assess his surroundings. Below in a tangle of treetops and second growth, he heard the faint sound of a deer snorting. Hurriedly, he walked toward them. When he reached the rim of the ravine, he heard deer running. He picked up his pace, drop-ping into the ravine for a closer look. Halfway down the steep incline, he solved the mystery. A breeding party was in progress.

At the bottom of the ravine, Bucky encountered several deer scattered among a thick tangle of blowdowns along a trickling stream. A yearling buck and two fawns were milling around a doe that was standing in a cluster of low-growing hemlock saplings, just above the stream bed. The doe was Buttercup. As Bucky approached, Buttercup turned to look into the thick brush that bordered the creek. The massive buck paused, realizing there was probably another deer nearby. Just then, a big 9-pointer came into view. As the 9-pointer trotted toward Buttercup, she bolted and ran for a tangle of treetops upstream and bedded. Buttons and Princess followed, but they had learned in the previous weeks that it was not safe to stay too close to their mother, so they bedded farther upstream. Immediately, Bucky took up the chase, trying to cut off the 9-pointer's approach to Buttercup. When the 9-pointer saw Bucky, he halted. He knew he was no match for the big buck. In a split-second, everything came to a halt. None of the deer dared move until they knew what Bucky's next move was.

When a doe enters estrus, there is no place to hide. Bucks will hound them until they are bred.

Finally, Bucky dropped his ears and aggressively snort-wheezed, dragging out the wheeze portion for several seconds. The 9-pointer's eyes drifted from the big buck to the doe bedded in the treetops. Before the 9-pointer could refocus, Bucky charged. The 9-pointer pivoted on his hind legs and ran, narrowly escaping Bucky's ice-pick tines. Bucky cut the chase off after several yards and returned to Buttercup. After milling around for a couple of minutes, the 9-pointer bedded within sight of the old doe. When the woods calmed, Bucky approached Buttercup and lay close to her so he could guard her from the lingering 9-pointer.

For the next hour, all was quiet, with only the sound of squirrels searching for acorns audible above the soft trickle of the stream. Then, amidst the jeering of blue jays, something caught Bucky's attention. He focused on a thick second-growth stand of hemlocks in the distance. A deer emerged and stood motionless. It was Skipper and a 7-point yearling buck. Buttercup was attracting a lot of attention. There were four bucks surrounding her. Only one would do the breeding, and Bucky was on a mission to make sure it was him. The others had different thoughts.

The small 7-pointer was so focused on Buttercup that he didn't see Bucky bedded near her. He moved with a gait that more resembled a dog moving to point a bird than that of a walking whitetail. As he approached, his neck was stretched out, with his nose barely off the ground. Buttercup just stared at him as the thudding of his hooves became louder. Bucky had seen enough. When the 7-pointer got within a few yards of Buttercup, Bucky sprang from his bed and snort-wheezed. Then, like a lightning bolt, Bucky charged the small buck. The 7-pointer never knew what hit him. The force of Bucky ramming his huge rack into the buck's flank knocked the little buck off his feet. With cat-like quickness, the 7-pointer jumped up and ran through the woods, with Bucky right on his tail. Fifty yards into the chase, the 7-pointer jumped over a 6-foot-high tangle of branches in a deadfall. The height of the deadfall made Bucky cut off the chase.

With Bucky preoccupied, the 2 ½-year-old 8-pointer jumped to his feet and started walking toward the bedded Buttercup. Skipper, feeling brave that another buck was leading the charge, followed closely behind. When the bucks neared the doe, Bucky came trotting back. The woods exploded. Skipper, aware of Bucky's temperament from past encounters, bolted first and

▲ Most fights between mature bucks during the rut are explosive. It's not uncommon for one or both of the combatants to sustain a serious injury during these fights.

◄ During violent fights a buck's antlers are used to inflict severe injury to one or both of the combatants.

quickly distanced himself from the action. The 8-pointer held his ground, and Bucky ran straight for him. Leaves flew as the 8-pointer reversed course. Just as Bucky was about to ram his antlers into the 8-pointer's flank, the smaller buck jumped a deadfall. Without breaking stride, the bucks flew over the downed tree's dead branches, like two steeplechasers clearing a water hazard. The bucks disappeared from sight, with Bucky in hot pursuit. When the woods calmed momentarily, Skipper charged Buttercup. Although Buttercup was his mother, the little 6-pointer couldn't contain himself. Nothing would stop him from trying to procreate.

The sight of Skipper charging toward her made Buttercup bolt from her sanctuary. In one motion, she was up and gone, frantically running northward. The entire time, Buttons and Princess had watched the action from a distance. When they saw their mother run, they took off to join her. Skipper found himself chasing three deer. It didn't take Bucky long to return. When he discovered Buttercup was gone, he quickly picked up her trail and took off to find her. She was easy to follow, and after 100 yards, Bucky found four deer standing in open hardwoods. The big buck pulled his ears back, lowered his head, bristled his fur and slowly walked toward Skipper in a stiff-legged fashion. Skipper was in trouble. After pausing for a second, he turned and ran as fast as he could toward a nearby clearcut.

With Skipper out of sight, Bucky turned on the fawns, which were standing near Buttercup. Buttons and Princess had seen enough of Bucky's aggressive actions to know that they had to distance themselves from their mother. In unison, they turned and ran a considerable distance before stopping to survey the situation.

Finally, Buttercup and Bucky were alone. The big buck, expecting the doe might run, slowly walked around a fallen oak tree next to her. Buttercup looked back over her shoulder at Bucky as he approached. She didn't move. Bucky smelled her and then began to delicately lick her flank. Slowly, the big buck slid his jaw onto Buttercup's back and mounted her. Within 30 seconds, she was bred. Bucky slid off Buttercup's back and stood next to her for a few moments. As he did, she looked at him from point-blank range.

Slowly, Buttercup walked several yards away from Bucky and began feeding on acorns. After consuming some, she bedded again. Bucky bedded next to her, and for the next several hours, he had Buttercup to himself. Twice during that period, the big doe rose from her bed, each time letting Bucky breed her.

With nightfall closing in, the bedded buck and doe were alerted by foot-falls in the leaves. Something was coming. Within seconds, a huge 10-pointer emerged from behind a big oak tree. The buck's body language made it obvious he could smell the estrous doe and was coming to crash Bucky's party.

When the big 10-pointer closed in, Bucky rose to his feet. The intruding buck stopped, shifting his gaze back and forth between Bucky and Buttercup. Bucky's short fuse was about to get shorter. For the next minute, Bucky let out a series of snort-wheezes and guttural grunts, which was his way of drawing a line in the sand. It was time to brawl. Bucky charged the 10-pointer, which made Buttercup bolt from her bed and scamper a short distance before stopping to witness the action.

During a doe's 24-hour estrus period, she might let one or more bucks breed her multiple times.

After rain or snow, bucks will be active, working primary scrapes to leave their scent.

Vital Information

• When November arrives in the North, a buck's testosterone and sperm count and a doe's estrogen level peak, setting the stage for breeding.
• Trail-camera data has shown that peak daylight buck activity during the rut occurs during the two hours after dawn.
• Sporadic rut behavior becomes the norm when intensive buck harvest leaves only yearling bucks for breeding. The higher the mature buck population – 3½-year-olds or older – the more intense the rut.
• When a doe nears her estrus cycle, she becomes more active.
• In the hours before estrus, does commonly deposit their scent on a scrape's overhanging branch and urinate into the bare ground beneath the branch.
• During November, flehmening, or lip-curling, peaks as bucks smell and analyze almost every patch of doe urine they encounter.
• A buck will stay with a doe for up to 72 hours during her estrus cycle, 12 to 24 hours before she enters estrus, 24 hours when she's in estrus and 12 to 24 hours after she cycles out of estrus.
• One or more bucks might breed a doe eight or more times during her 24-hour estrus cycle.
• Studies have shown that a buck can pick up the scent of an estrous doe from more than 300 yards.
• When bucks vie for breeding rights to an estrous doe, aggressive fights occur to determine which buck does the breeding. In some cases, a buck dies.

By late November, bucks are exhausted from the rut. With breeding complete, they spend most of the day bedded.

panting so hard that it looked like steam was shooting from their mouths when they exhaled. After pausing, the bucks resumed the struggle. Both deer were nearly exhausted. Bucky gained leverage and pushed the 10-pointer against a fallen log. The force made their antlers unlock. Before the 10-pointer could regain his balance, Bucky rammed his antlers into the 10-pointer's side. On impact, the 10-pointer let out a loud moan. Before the sound faded, Bucky rammed his rack into the buck's flank. The force prompted the 10-pointer to spring to his feet and half-stumble, half-run through the woods. The fight was finished.

For the next hour, Bucky stood awkwardly in the forest, trying to recover from the fight. Steam shot from his mouth and nostrils with every breath. He couldn't stop panting. When nightfall swallowed the forest, Bucky had recovered enough to look for Buttercup. He found her bedded nearby.

Throughout the night, Bucky ran off several bucks that tried to take Buttercup from him. Though none were big enough, each encounter forced him to burn energy he didn't have. Three more times during the night, Bucky bred Buttercup before she cycled out of estrus. The next morning, with the smell of estrus no longer present, Bucky left Buttercup and began walking. Although he was exhausted, hungry and banged up from fighting, he knew his mission was not complete until every doe in his territory was bred.

The bucks' antlers came together with tremendous force, and the woods turned into a war zone. First the 10-pointer and then Bucky gained the advantage as they pushed each other back and forth. Periodically, they unlocked and relocked their antlers, after which they would resume pushing. Throughout the ordeal, small saplings in their path were flattened or ripped from the ground. The sounds of antlers clacking together, branches breaking, and moaning and grunting could be heard everywhere. It was a fight to the death.

Little more than a minute into the fight, the bucks' antlers locked tightly together. They couldn't separate. With their rumps raised and chins resting on the ground, the bucks tried to catch their breath. They were

CHAPTER 9

December

Fluffy quarter-sized snowflakes floated from the sky as Bucky and Skipper greeted the new day. They had bedded within sight of each other in a thick, swampy area after they had fed on acorns throughout the night. With the rut finished, the bucks were trying to recover from their November marathon. Though Skipper had done very little breeding during the rut, he had depleted most of his fat reserves going from one breeding party to another. Bucky was not so fortunate. When October gave way to November, the big buck's body was plump and fat, weighing about 250 pounds. But with winter knocking at the door, Bucky was 50 pounds lighter and desperately needed to replace some of his fat reserves. Winter would not be kind to any deer that entered winter physically stressed. An hour after first light, with light snow still floating from the sky, two other bucks entered the swamp and bedded close to Bucky and Skipper. The spike and a 3½-year-old 8-pointer were bucks Bucky and Skipper knew but had not seen since October. By midmorning, snow began covering the bare ground and each buck's back. They made no attempt to acknowledge each other as they burned daylight dozing and chewing their cud. Their mission was to recover from the rut, and the best way they could do so was to eat and sleep.

Bucky nuzzled his nose between his hind legs and closed his eyes. For the next hour, he drifted in and out of a deep sleep. Periodically, his ears twitched and changed position when he heard the familiar "fee-bee, fee-bee, fee-bee" of chickadees calling each another. Even while sleeping, he deciphered every sound in the forest. Nothing got past him.

About midday, the bucks were jolted from their slumber by the constant jeering of scolding blue jays. Though the noise was far away, the deer took notice because something was obviously bothering the birds. The squawking didn't last long, and with the scene calm, the bucks resumed grooming and dozing. Minutes later, Skipper thought he heard a branch snap at the edge of the swamp. The freshly fallen snow had begun accumulating on the willow and alder branches, making it difficult to see far. When a second branch broke, the bucks heard the sound and immediately trained their eyes toward it. Something was coming.

Skipper, who was closest to the sound, stood. A man was coming. Skipper snorted and began running toward an opening in the swamp that led to an open

▲ Does that were not bred in November cycle into estrus in December. When that occurs, multiple bucks form breeding parties and vie for breeding rights. This adds to their already overstressed condition.

field. Quickly, Bucky and the other two bucks bolted from their beds and followed. After maneuvering his huge antlers through the thick alders, Bucky broke into the open and began bounding across a narrow field. He saw Skipper and the other bucks running ahead of him. Boom! Boom! Boom! When the second boom from the hunter's rifle rang out, the 8-pointer ahead of Skipper fell, so Bucky and Skipper quickly changed course. Within seconds, they crossed the field and bounded into a standing cornfield. The bucks were safe for the moment but had to distance themselves from the hunters. They wasted no time moving between the corn rows to the end of the field. Just before leaving the field, the bucks stopped to assess the situation. For several seconds, they tested the wind for danger. Nothing. Quickly, they trotted out of the field into a thick second-growth stand of timber. With the snow falling heavily, the bucks knew it would be difficult for the hunters to track them. Slowly, they worked their way to the thickest part of the forest and bedded for the remaining daylight hours.

▲ When November has passed and most does have been bred, bucks will have lost up to 20 percent of their pre-rut body weight. To survive the coming winter, bucks feed heavily to regain weight.

▲ When breeding occurs in December, does are stressed by harassment from bucks.

Buttercup, Daisy and the fawns were glad to see December arrive, for it meant the rut was finished. They wouldn't have to contend with being harassed by every buck in their territory. With their world returned to normal, the does spent their days bedding in their favorite oak flat, where they only had to stand and eat the acorns that carpeted the ground. Life was good.

Dawn arrived clear and cold, with heavy hoar frost clinging to everything. For more than two hours, Buttercup and her family had been bedded comfortably amongst a stand of mature oak trees after a night of feeding in a nearby cornfield. It was a perfect place to spend the day. Near the cornfield, the deer could hear a large machine working — a combine harvesting corn.

Not long after sunrise, Buttons stood and stretched. He was somewhat high-strung and was never able to bed as long as Buttercup, Daisy or Princess. After stretching, he let out a soft grunt and walked to his mother. Buttercup sensed he was looking for affection, so she stretched out her head and neck. The button buck nuzzled his head against Buttercup's and began to lick his mother's ear and neck. For the next five minutes, the bedded doe and fawn groomed each other. Their love fest ended when a pileated woodpecker landed on the trunk of a nearby dead tree and began hammering away. For several seconds, Buttons watched the noisy bird as it pounded pieces of the dead tree with its beak. Intrigued by the flying chips, Buttons walked toward the woodpecker. It ignored the fawn's approach. When Buttons got to within 10 feet of the bird, it stopped pecking and looked intently at the fawn. Seconds later, the woodpecker flew away, with Buttons looking inquisitively at the scattered shavings on the ground.

Near midday, with the hum of the machine echoing through the forest, a flock of turkeys worked their way toward the bedded does and fawns. Minutes passed as Buttercup, Daisy, Buttons and Princess watched the turkeys scratch beneath the blanket of leaves, looking for acorns. Soon, 15 turkeys were mingling around the bedded deer. Now and then, a turkey would stop to look at the does as they rested. Otherwise, they paid little attention to Buttercup and her clan.

Soon, the biggest hen in the group walked to within a few feet of Princess and began scratching in the leaves. Each time she cuffed at the leaves, debris flew into the air. That scared Princess, because some of the leaves landed on her, making her uncomfortable. When the hen continued pelting her with leaves and sticks, Princess stood

▼ In the North, a small percentage of doe fawns will be bred in December.

Even when December's snow arrives, bucks are still driven to make rubs, which will continue until their antlers are cast.

and looked the hen in the eye at point-blank range. When the hen didn't back down, Princess walked to her mother and bedded.

Long shadows formed around the bedded does and fawns as the sun set. Other than the woodpecker and turkey incidents, the day had been uneventful. With light fading and the temperature dropping rapidly, Buttercup, Daisy, Buttons and Princess rose from their beds and began to feed. The acorns tasted good, but they knew the cornfield's bounty would taste better. At nightfall, they began their trek to the cornfield.

When they reached the cornfield, the deer were greeted by a new landscape. Before them sat a huge machine, parked in the middle of broken-down cornstalks. They were surprised to see that half of their favorite feeding location had disappeared since the previous night. At the far end of the field, the group saw other deer feeding in the moonlight. Slowly, the four walked toward the standing portion of the field. Soon, Buttercup, Daisy, Buttons and Princess were busy knocking down stalks and peeling the leaves off ears of corn. It would be the last night they'd be able to gorge themselves on the field's corn crop.

It had been several weeks since the rut had trickled to an end, and Bucky was beginning to regain some of the weight he'd lost in November. With only a few fields of corn still standing, the big buck and other deer had shifted their feeding habits from the corn they loved to acorns and turnips. Significant snowfall had not yet arrived, and with the ground not frozen, two big turnip fields in Bucky and Skipper's territory received a lot of attention from deer.

▲ To conserve energy, whitetails bed extensively when the cold and snow of winter arrive.

Since the encounter with hunters, Skipper had noticed that Bucky was accepting him more. As a result, the bucks formed their own little bachelor group. In November, they had been fierce competitors, but now they were becoming buddies.

Daylight was at a premium, as the shortest day of the year approached. The day had been clear, windless and very cold, even for December. Bucky and Skipper had spent the day bedded in a dense thicket 100 yards from a large turnip field. As the sun set, they rose from their beds, stretched and urinated. For several minutes, they groomed themselves, all the while smelling the thermals and listening intently for danger. Though the day had been sunny, they sensed a big storm approaching and knew they must feed before it arrived.

With darkness settling in, the bucks slowly walked to the field. Throughout the night Bucky, Skipper and many other deer fed heavily on turnip greens and the nutrient-rich tubers they pried from the ground. A few hours before dawn, clouds began to hide the stars, and wind began blowing. With their bellies full, Bucky and Skipper left the field an hour before dawn and walked toward a spruce plantation that would provide cover from whatever nature was sending their way.

Although the rut is finished, a whitetail's quest for dominance doesn't end. Here, an older mature buck attempts to exert force on a younger mature buck.

▲ Exhausted from the rut, bucks re-form bachelor groups and rest extensively to conserve energy lost during the previous month.

Vital Information

• By early December, a buck's testosterone level and sperm count have dropped to half of what they were in early November, so his urge to breed is less intense. He feeds more frequently to replenish the body fat he lost during the rut.

• In areas of the North where doe fawns reach 80 pounds by Dec. 1, about 15 percent will breed in December.

• Northern trail-camera data has shown that peak daytime deer activity in December occurs during the final two hours of daylight. The second greatest period of daylight activity occurs from 10 a.m. to 1 p.m.

• Bucks and does will bed up to 75 percent of the time in December. Typically, they bed for two to four hours before getting up and moving.

• In areas where winter can be harsh, it's critical for whitetails to enter winter with a 90-day supply of fat on their bodies. Anything less puts them at risk of starvation if winter is cold and snowy.

• Whitetails will consume seven or more pounds of browse per day if it's available. Though low in nutrition, browse can sustain a deer through winter.

• If a Northern fawn enters winter weighing less than 70 pounds, its chances of surviving a harsh winter decrease dramatically.

• When the Europeans came to North America, there were about 20 million whitetails on the continent. Today, the population is estimated at more than 30 million. Without the controlled harvest through legal hunting, widespread deer starvation would occur.

Soon after daybreak, the wind died, and it started to sleet. Although Bucky and Skipper were bedded under the spruce canopy, ice droplets found their way to the deer's bodies. An hour into the storm, the sleet turned to snow, and the wind picked up again. During the next eight hours, the spruce plantation was transformed into a winter wonderland, with spruce bows bent to the breaking point with ice and snow. Throughout the storm, neither Skipper nor Bucky moved from their beds. They just stayed there, chewing their cuds and dozing. Occasionally, they groomed themselves between naps.

During the day, more than six inches of snow fell, and the storm's heavy cloud cover brought darkness earlier than normal. Soon after nightfall, the snow stopped falling, and the wind calmed. Hungry, Skipper rose from his bed and stretched, and a coating of snow slid off of his back like an avalanche cascading down a mountain. More than an inch of snow covered his neck and parts of his back. He hunched up, arching his back like a rainbow, and then shook every inch of his body. Snow jettisoned in every direction. From his bed, Bucky watched what Skipper had done, and then the big buck stood and followed suit. It was time to look for food.

CHAPTER 10

January

It had been days since Bucky had seen Skipper. Alone, the big buck had taken up residence in a south-facing spruce plantation. His years of surviving harsh Northern winters had taught him that conifer forests were much warmer than open hardwoods when January put a lock on his world. He also knew that south-facing slopes let him take advantage of any sunlight winter might parcel out.

As with so many preceding days, Bucky greeted the slate-gray day of winter bedded in a foot of snow under a canopy of spruce boughs. Snow had been falling since before daylight, and the big buck was content to stay curled up with his nose tucked between his hind legs. It was the best way to stay warm and conserve energy in the frigid elements. Aside from the periodic chatter of red squirrels feeding on spruce cones high in the trees, and the frequent flutter of chickadees landing on branches around him, the forest was quiet.

As midday approached, snowflakes stopped falling, and the sky brightened. Breaks in the clouds let the sun's rays shaft through the branches and touch Bucky's body. Within minutes, he felt his body temperature increase. It had been several days since the sun had shone, so the sunlight made the buck want to look for food. Slowly, Bucky stood and looked around. Six inches of fluffy snow made it impossible to see far. Fortunately, his nose told him he was free from danger.

▼ Grooming is a bonding behavior exhibited by bucks and does.

▼ Though the rut is a faint memory in January, bucks still spar with each other.

When whitetails rise from their beds in winter, one of the first things they do is shake any accumulated snow from their body.

Browse makes up most of a deer's winter diet. Most browse at this time is well below the nutritional levels deer require. That's why it's critical for deer to enter January with adequate fat reserves. In regions without adequate nutrition, bucks begin casting their antlers in early January. Look for sheds near bedding areas, food sources and south–facing locations.

Caked with snow, the big buck hunched up and shook his body, sending a cloud of snow in every direction. Although he hadn't eaten for two days, he didn't feel particularly hungry, but knew he must find food to survive. After grooming himself, he began walking toward a thick stand of honeysuckle that bordered the spruce plantation. If any leaves were still hanging on the branches, they would provide enough nourishment until he could make it to his favorite cornfield.

Within minutes, Bucky entered the thick tangle of honeysuckle and began to feed on the few leaves still clinging to the bushes. The sun's rays felt good and warmed his body as he fed. Each time he nipped off another leaf, snow fell on his face, neck and antlers, making him look more like a white-masked bandit than a majestic buck. As he was about to reach for another leaf, he saw movement to his left. Slowly, two deer fed toward the snow-covered buck. They were fawns, and judging from their tattered appearance, they were not coping very well in the frigid conditions.

A fawn attempts to communicate and bond with a two-year-old buck.

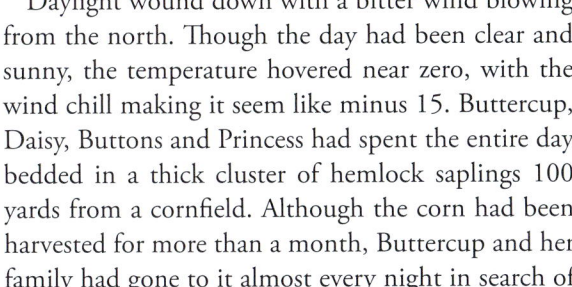

▲ Whitetails heavily browse goldenrod leaves during winter. Though marginal from a nutritional standpoint, the leaves help sustain deer through the winter.

During the hour the fawns fed nearby, Bucky never saw their mother. Before long, the deer fed from sight, leaving him alone again. After feeding, Bucky walked to an open slope that bordered the spruce and honeysuckle. Wanting to take in the sun's warmth, the old buck bedded next to a pine tree that protected him from the wind. Though the day was bitterly cold, the direct rays from the sun made his bedding location feel 20 degrees warmer.

▲ If any corn is standing in January, every deer in the area will gravitate to it.

Daylight wound down with a bitter wind blowing from the north. Though the day had been clear and sunny, the temperature hovered near zero, with the wind chill making it seem like minus 15. Buttercup, Daisy, Buttons and Princess had spent the entire day bedded in a thick cluster of hemlock saplings 100 yards from a cornfield. Although the corn had been harvested for more than a month, Buttercup and her family had gone to it almost every night in search of the kernels the combine left behind.

As the last rays of light reflected a reddish glow off the snow, Buttercup sank to her knees in snow as she stepped into the field. Slowly, she and the other three deer walked past one hole after another in the snow. Each bulldozed hole showed where deer had pawed the snow away to bare ground, looking for corn. Slowly, the four sniffed the top of the snow as they tried to smell for any kernels that might be on the ground. Finally, Buttercup and Daisy stopped and began to cuff away the snow. For the next several minutes, Buttons and Princess stood and watched as their mother and sister shoveled away snow with their hoofs. Though the process was a huge energy drain, it was the only way for the deer to access the corn.

It didn't take them long to reach bare ground and the partially broken cobs of corn. The sound of Buttercup and Daisy chewing on the kernels brought Buttons and Princess to the sides of their mother and sister. They hoped to share in what little food was available. Unfortunately, when the fawns wedged their noses into the snow crater, Buttercup and Daisy had already eaten every kernel. The fawns were quickly learning that if they wanted to survive, they needed to fend for themselves, which meant digging for their own food.

During the next three hours, Buttercup and her family were joined by five other deer hoping to find enough corn to sustain them. As they cuffed holes in the snow, a pack of coyotes howled and yipped in the distance. The sound was a grim reminder of the dangers deer faced during the long January nights. It proved to be a difficult night, and by dawn, Buttercup, Daisy, Buttons and Princess were exhausted from digging in the snow. They were also still very hungry.

As the deer left the cornfield and headed back to their hemlock thicket, they stopped every few yards to nibble on saplings exposed just above the snow line. The wind that howled throughout the night had subsided, and with the woods dead calm, the sound of the browse being snipped off by the molar teeth of the deer could be easily heard. When the deer neared the hemlock cover, they stepped onto a snow-packed trail and walked single file into the thicket before settling in the same depressions each had bedded in for several consecutive days.

During January, it's common for whitetails to bed up to 90 percent of the time, especially during inclement weather.

Although the rut is finished, bucks still lip curl when they encounter doe urine.

▲ Browse makes up most of a whitetail's diet in winter. Even the best browse provides food with only marginal protein levels.

▲ Though rare, some does cycle into estrus in January and are bred. When this happens, nearly every buck in the area gets involved until the breeding is concluded.

Dawn found Skipper and two other bucks milling around in an old apple orchard that bordered a gravel country road. With the temperature hovering near freezing, the bucks hoped to find a few frozen apples to eat before heading to their daytime beds. As thermals sifted through the orchard, the smell of apples made Skipper pause and look skyward. Spotting apples hanging on branches overhead, the young buck stretched out his neck and tried to reach one. As hard as he tried, he couldn't get within two feet of the lowest hanging apple. Carefully, he positioned his feet in the foot-deep snow and thrust his body upward. Just as he was about to secure the apple in his mouth, he lost his balance and returned to the ground. Gathering himself, he tried again and failed again. Finally, on the third try, Skipper catapulted himself into the air and grabbed the apple with his mouth. As he descended, the apple slipped from his grasp and fell into the snow. Without hesitation, the young buck burrowed his nose into the deep snow, sniffing around until he found it. With his face covered in snow, Skipper quickly shifted the apple to the rear of his mouth and began gumming it with his sharp molar teeth. No matter how hard he tried, he couldn't penetrate the apple's outer shell. It was like trying to eat a round ice cube. After a couple of minutes, the apple thawed to the point that Skipper could chew and swallow it. Although it was a far cry from what it had been in October, the apple was still a treat.

For the next two hours, Skipper and the other deer fed on goldenrod leaves. Although they weren't as tasty as apples, the tiny leaves were one of the bucks' favorite wintertime foods. Around midday, the skies darkened, and it began to snow. The bucks continued to feed on goldenrod leaves, hoping to fill their bellies before the weather forced them to bed. It was not to be. Discouraged by the heavy snowfall, the bucks gave up and headed toward the road and the woods on the other side. It was there that the foursome hoped to bed during the storm.

Before Skipper could reach the road, he paused. A vehicle was fast approaching. The two bucks ahead of him bolted for the road. As Skipper watched, the first buck cleared the gravel strip with one bound. The second was not so lucky. When the buck reached

the road, the vehicle hit him broadside, hurtling him into the ditch. As the vehicle skidded sideways down the snow-covered road, Skipper panicked and ran across the highway. He never stopped running until he reached the woods. Safe, Skipper paused inside the woods line to see where the other bucks were. Snow was falling so hard that he couldn't see anything. It was a near white-out. Feeling alone, Skipper quickly walked to the edge of a nearby ridge and descended into a deep conifer-choked ravine. It was time to bed till the storm passed.

The sound of wingbeats jolted Bucky from his sleep. A crow had flown right over his head as he lay curled in a cluster of beech saplings, sleeping in the deep snow. The scavenger had seen Bucky lying motionless, and thinking he was dead, came in looking for an easy meal. When he saw that the big buck was alive, the crow landed in a nearby beech tree to survey the situation. Quietly, the crow looked around for several minutes before cawing several times and then flying off.

For the next two hours, Bucky remained still. His warm body had melted the snow in his bed enough that only his neck, head and antlers were visible above the snow line. Big chunks of ice and snow gripped his antlers, and the air temperature was so cold that frost had formed on his eyelashes and whiskers. Soon after nightfall, the full moon rose, casting eerie shadows in the forest. The night was calm, letting every sound of the forest echo for a great distance. Bucky had been nestled in his snow bunker for

Here, a yearling spike buck spars with a bedded mature buck to pass the long, dreary days of winter.

nearly 12 hours. It was time to get up and move. He needed to move so he could increase his body temperature and also needed to eat. With a full moon showing the way, he stood, stretched and then walked a few feet from his bed to urinate. For several moments, he paused to groom his shoulder and back before slowly walking through the open hardwoods. Less than a mile away was a standing cornfield, and Bucky was ready for a decent meal.

Bucky knew deer were in the field before he reached it. When he got close, the sound of leaves being husked from the corn cobs immediately told him deer were feeding. When he reached the field, the first deer he encountered was Skipper. He hadn't seen the young buck for a long time. Although the temperature was well below zero, the bucks spent time bonding as they fed and groomed each other.

Each time Bucky picked up his head from feeding on a corn cob he had wedged between his feet, Skipper tried to get him to spar. After rejecting the young buck's tease several times, Bucky swung his heavy antlers sideways, and Skipper took the bait. For the next several minutes, the two clicked and clacked their antlers together under the frigid January full moon.

Vital Information
- During harsh Northern winters, deer bed up to 90 percent of the time to conserve energy.
- Deer seek bedding areas that offer thick overhead canopies, such as cedars, spruce and hemlocks out of the wind. Such locations can be 10 degrees warmer than open hardwoods.
- South-facing exposures are prime locations to find deer from January through March. Whitetails move to these areas to take advantage of the sun.
- To conserve energy, deer might go two or more days without eating if conditions are harsh.
- Browse makes up most of a deer's winter diet. Most browse at this time is less than five percent protein and high in fiber, well below the nutritional levels deer require. That's why it's critical for deer to enter January with adequate fat reserves.
- In regions without adequate nutrition, bucks begin casting their antlers in early January. Look for sheds in or near bedding areas, food sources and south-facing locations.

CHAPTER 11

February

Dawn was magical. As the sun inched toward the horizon, its rays reflected off the oncoming cloud bank, painting it varying shades of red and gold. God was showing his majesty this cold winter morning. Unfortunately, dawn's light show would be the high point of the day. Afterward, everything would go downhill.

During the night, Buttercup, Daisy, Buttons and Princess had tried to dig for what little corn was left in the snow-covered field. Their efforts were in vain, because they burned more energy than they received from the few kernels of corn they found. Hungry and exhausted from the grind of winter, they returned to their hemlock sanctuary well before dawn to bed. It had been more than 40 days since they had consumed any quality food, and it was beginning to show on the fawns. The fat deposits Buttons and Princess had built up during autumn were nearly depleted. They desperately needed a break from Mother Nature.

Soon after dawn's incredible light show ended, snow began to fall from the slate-gray sky. The wind picked up, driving the snowflakes sideways through the cluster of hemlocks in which the deer were bedded. By mid-day the wind-driven snow was falling at three inches per hour, with no sign of letting up. Although Buttons and Princess were bedded less than 10 yards from Buttercup, there were times they couldn't see each other because of the snow swirling in the hemlock thicket. It was an all-out blizzard.

Buttercup and her family curled up in their beds, trying to brave the elements. By early afternoon, snow covered the bedded deer. Only the brown of their eyes differentiated them from surrounding snow-covered logs. Late in the afternoon, with heavy snow still falling and swirling, Buttercup rose from her bed so she could urinate. She had nearly three inches of snow caked on her body. After shaking the snow off, she walked to Princess and looked into her eyes. Princess just looked at her mother through the snow mask she was wearing. Buttercup turned her attention to Buttons, who was bedded a few feet from Princess. Only his snow-

▼ Winter in the North can be torturous to whitetails. In lake-effect areas, more than 200 inches of snow often falls by late February.

▼ When weather is cold and snowy, deer might bed for days to conserve energy.

▲ Activity in a wintering bachelor group increases when there's a break in the weather.

covered head was sticking above the snow line. For a few seconds, Buttercup and Buttons just stared at each other. He had never been through a storm like this and didn't know what to do, so he stayed bedded.

When Daisy saw that her mother was standing next to the fawns, she got up from her bed, shook the snow off her body and slowly walked toward Buttercup. For the next few minutes, the does stood in the snowy fog, trying to groom each other, with Buttons and Princess watching. The cold, wind-driven snow proved to be too much for the does, and within minutes, they returned to their beds.

More than two feet of snow had fallen since dawn, and with the blizzard still raging, daylight vanished. Throughout the night, the does and fawns remained bedded in the hemlocks. Several times during the night, they heard trees snapping in the gale-force wind. Although Buttercup had been through other blizzards, this one was as intense as she had ever seen. Soon after dawn, the snow stopped falling, and the wind died. It was still overcast, but at least the storm had ended. In the distance, Buttercup heard a snowplow as it cleared a gravel roadway.

▲ Whitetails might consume up to five pounds of browse each day if it's available.

Almost three feet of new snow covered the ground. Buttercup and Daisy tried to venture away from the hemlock canopy but quickly returned. The snow was chest deep, making it impossible for them to move. The fawns, not knowing what to make of the storm, rose from their beds. After shaking snow off their coats, they also attempted to walk a few yards toward open hardwoods. They quickly gave up because the snow was too deep. Unable to navigate, Buttercup and her family stayed put until the snow settled enough for them to move around.

❧

The long winter was taking a toll on Bucky. For more than two months, he had tried to find enough browse to eat, but with each day, the struggle was becoming more difficult. The blizzard only added to his misery.

Bucky weathered the storm by bedding in a tangle of treetops and evergreen saplings in the bottom of a deep ravine. Although the location protected him from the harsh wind, he couldn't escape the snowfall. When the blizzard ended, he was covered with a thick coat of snow, with only his eyes, ears and antlers visible above the snow pack. With the snow finished, Bucky's first action was to shed the snow on his body. After that, he needed to find food.

▼ If the deer herd is healthy, most bucks will cast their antlers in February. This buck cast his antler Feb. 23.

Within seconds of standing, Bucky shook off the thick layer. As he jerked his head from side to side, something strange happened. His left antler fell off his head and stuck in the snow beside him. Startled, he lunged backward and stared at it for a few moments, and then leaned forward and smelled the bloody base of the antler. On his forehead, where the antler had been attached to his skull, droplets of blood began to ooze and drip down the side of his face. Bucky panicked at the smell and sight of blood. Unable to walk in the chest-deep snow, he took several bounds before stopping to look back. His head felt different than it had minutes before. It felt lopsided. Unsure what was happening, he shook his head from side to side. Then the right antler separated from his skull and fell into the snow beside him. Though his head felt lighter, he still smelled blood. Frantically, he bounded off through the thick brush before stopping 50 yards away. When he stopped, he couldn't see the hard antlers he had carried the past six months. His weapons were gone, and with it his identity. Now, at least from a distance, he looked like every other deer.

Moments after casting his antler, the buck smells a doe's urine and lip curls.

In northern regions with good nutrition, most bucks will cast their antlers between late January and mid–February.

Slowly, the big buck navigated through the deep snow until he reached a spot that was being logged. Everywhere he looked, there were treetops at awkward angles on the forest floor. To Bucky, this was a godsend, because the browse on the tops provided food he desperately needed. For the next hour, Bucky meandered through the tangle of oak and maple tops, browsing on the ends of every tree branch he could reach. Just as he was about to leave, Skipper and two other deer showed up to feed on the tops. Unlike Bucky, Skipper had shed only one of his antlers, but even without complete racks, the bucks knew each other.

At one point, the bucks browsed on the same treetop. Skipper noticed the blood dripping down

▲ Bucks seldom cast both antlers at the same time. Although this buck has cast one antler, he's still able to rub without the other falling off.

▶ A buck that has recently cast one antler steps over a snowy log as he searches for food.

the side of Bucky's head and began sniffing the air near the big buck's forehead. Cautiously, Skipper stretched out his neck and licked the blood from the side of Bucky's face. Within seconds, the blood was gone. When Skipper tried to position himself to lick the other side, Bucky showed his dominance by raising his head into the air. Although he no longer had antlers, Bucky wanted to make sure Skipper knew who was boss. Skipper slowly backed up, turned around, walked to another treetop and resumed browsing.

Feeling full, Bucky turned and began walking in the deep snow. He didn't go far before stopping under a big pine tree. For the next minute, he pawed and cuffed snow away as he attempted to make a bed. Unable to reach bare ground, Bucky stopped pawing, dropped to his knees and eased his body into the snow. He needed to rest for a while.

From his bed, Bucky noticed that several new deer had come to feed on the freshly cut treetops. Not far from him, Skipper and an 8-pointer were browsing on branches that hung from a huge oak top. Within minutes, the two paused from browsing and began to spar. With their bellies touching the snow, the bucks aggressively raked their antlers together several times before separating and walking in opposite directions.

Just as Skipper was about leave the cut-over, every deer became alert. A big doe feeding in the tops had spotted movement in the woods above the cut area. With her tail flared to warn of danger, the doe stood erect. Something was coming. Before long, four deer emerged from the sea of white. It was Buttercup, Daisy, Buttons and Princess, who had come to feed. When they were in sight, the other deer calmed and resumed browsing.

Buttercup, Daisy, Buttons and Princess waded through the snow and began browsing. Each stem of browse tasted good, especially after not eating anything the previous day. For the next hour, the foursome fed as other deer came and went. With daylight fading,

▼ Although he's already cast his antlers, this mature buck attempts to show his dominance to another buck.

and their bellies full of maple and oak browse, Buttercup and her family poked past the bedded Bucky. It was time to bed for the night.

As nightfall settled in, the skies cleared, and before long, the star-studded heaven cast a dull glow in the forest. With only the stars to guide his way, Bucky got up from his bed, where he had been since early afternoon. It was time to eat again. For several seconds, he stood motionless, with only his ears moving back and forth, monitoring the surroundings for danger. Sensing all was clear, the buck arched his body and stretched, leaning forward. He then stepped in front of his snow bunker and urinated before walking into the tangle of treetops to feed.

Soon after nightfall, the feeding deer were treated to "hoo-ah, hoo-hoo-hoo, hoo-hoo-hoo-hoo." A great horned owl was serenading the forest with his calling. As eerie as the sound was to Bucky and the other deer, it was far better than the sounds generated by the howling blizzard they had just endured.

A few hours before dawn, with temperatures well below freezing, a pack of coyotes began to yip and howl. The canine chorus made Bucky stop browsing and pay attention. Confident the coyotes were farther away than they sounded, he resumed browsing. He and every other deer were deathly afraid of coyotes, so whenever the dogs sang, deer took notice.

Bucky sensed that each day brought more daylight. Each day also brought more bird life to the forest.

A doe smells a rare find: two freshly cast antlers in the snow.

Vital Information
• In northern regions with good nutrition, most bucks will cast their antlers between late January and mid-February.
• Research has revealed that serious winter deer loss begins when there are 60 or more days with 15 inches of accumulated snow or 50 or more days with 20 inches.
• Mortality from coyotes and domestic dogs can be significant in areas with heavy snowfall.
• When winter is severe, it's common for fawns to make up more than 80 percent of winter mortality.
• If winter conditions are harsh, deer tend to stay in prime bedding areas, even if adequate food is less than a half-mile away.
• A Northern whitetail's monthly food consumption from January through March is about 30 percent what it is during other months.

Unlike January, when brutally cold temperatures and heavy snowpack made the forest very quiet, the end of February brought an awakening. By the end of the month, wildlife knew that spring was around the corner.

An hour before dawn, Bucky lay in his bed with his head resting on his side. Despite the severe cold, he slept comfortably. Unfortunately, with March about to burst onto the scene, his hope of sleeping soundly past dawn was out of the question, thanks to the likes of chickadees, blue jays and cardinals. It seemed as if every bird in the forest felt it had to usher in a new day.

Only 24 hours removed from the harsh blizzard, faint chirps began resonating through the crystal-clear air at dawn's early light. As the forest brightened, the "fee-bee, fee-bee, fee-bee" of chickadees, the "toolool, toolool, toolool" of blue jays and the "cheer, cheer, cheer" of a Northern cardinal got Bucky's attention. Unable to sleep through the birds' concert, Bucky rose from his bed, stretched and greeted the new day.

CHAPTER 12

March

Throughout the afternoon, light drizzle fell, making the 6-inch snowpack a blend of snow and slush. As night approached, eerie pockets of fog formed and drifted across the landscape. The temperature hovered near freezing, and light rain soon replaced the drizzle. Bucky and Skipper were bedded 20 yards apart in a cluster of fallen treetops at the head of a ravine, about 100 yards from the edge of a big snow-covered field.

As the temperature dropped to the freezing mark, ice formed on everything touched by rain. Tree branches sagged, and goldenrod stems collapsed under the weight. Despite the conditions, Bucky seemed unfazed by what was occurring. Skipper, however, had never been through an ice storm.

A buck that has already cast its antlers grooms a hard-antlered mature buck.

By midnight, Bucky wasn't sure he wanted to stay where he was, and Skipper was becoming increasingly scared. As the night wore on, conditions continued to deteriorate. Trees groaned and creaked under the weight of the ice. Now and then, the bucks heard a loud snap, followed by a tree crashing to the ground. Despite the branches breaking and trees falling, Bucky and Skipper stayed in their beds. It wasn't until dawn that they saw what had occurred during the night. Everything was covered with a thick coating of ice. No matter where they looked, visibility was limited to a few yards because of bent-over tree branches. The forest was a winter wonderland.

Soon after daybreak, the rain stopped, and a light southwest breeze began to blow. Minutes later, the wind intensified, causing the ice-bound trees to sway and creak. Each time the wind gusted, branches snapped under the weight of the ice and fell noisily to the ground, throwing shards of ice in every direction. The forest had turned into a war zone.

When a big pine tree near the bedded bucks buckled and roared to the ground, Bucky jumped to his feet

▼ Gaunt and hungry, these does and fawn spend their days searching for food. By March, most whitetails are stressed to the limit.

▼ Rutting behavior never ends for some bucks. Though antlerless, this buck takes a break from the grind of winter to check out the previous fall's rub.

and bolted for the open field. Having experienced several ice storms, he knew he must get to an opening if he expected to survive. When Skipper saw Bucky bolt from his bed, he wasted no time following. It only took the bucks seconds to reach the forest edge. Bucky didn't like what he found when he entered the field. A thick layer of ice had formed atop the snow. With each step, he broke through the crust, sending pieces of ice sliding across the icy surface. Bucky felt helpless. He knew he couldn't go back in the woods, but he couldn't stay in the open, either. The big buck wasted no time. With Skipper in tow, he began to walk across the field. On the far side was a clearcut with dense hardwood saplings, multiflora rose bushes and berry briars where he would be safe until the ice melted. It wasn't easy navigating the field. With each step the deer took, the blanket of ice cut into their ankles. When Bucky and Skipper finally reached the clearcut, they stopped to check for danger. Feeling safe, they walked a short distance into the dense maze of ice-covered branches and bedded.

◄ This antlerless buck shows he still has enough energy to jump a snowy log.

▼ A mature doe nourishes herself with berries from a multiflora rose bush. By March, any food is welcome.

By midday, the temperature increased enough to begin melting the ice. Water droplets rained on the bedded bucks, making them miserable. Though soaked, Bucky didn't dare move until he was sure it was safe to do so. By late afternoon, with most of the ice melted from the bushes, Bucky rose from his bed and looked around. Seconds later, he shook the water from his coat, spraying every nearby bush. It didn't take Skipper long to follow Bucky's lead. The little buck rose from his bed and shook the water off his body. Moments later, he poked between the wet bushes toward Bucky. Though the ice was gone from the bushes, the snow pack still had a thick layer of ice covering it, making walking noisy and difficult. The young buck walked to Bucky and stood next to him. When Bucky was unresponsive, Skipper began licking the big buck's neck and ears. For the next several minutes, the bucks took turns grooming each other before beginning to browse on the nearby oak and maple saplings. Rather than risk injuring themselves on the sharp ice that covered the snow, the bucks stayed in the clearcut until conditions improved.

▲ A doe attempts to groom her fawn during a March snowstorm. By this time, deer have been dealing with cold, snowy conditions for more than 100 days.

▲ Facing minimal snow depth, this doe group tries to feed in a frozen clover field. Usually, deer burn more energy than they consume from the frozen clover leaves.

In snow-belt regions, it's common for starvation to occur if a deer loses 30 percent of its pre-rut body weight during winter. If winter conditions are normal, a doe will emerge from winter in better physical shape than bucks who entered winter worn down from the rut.

Two weeks had passed since the ice storm transformed Buttercup's world into a clutter of fallen branches and downed trees. With many of her woodland trails blocked with debris from the storm, she found it difficult to travel between her favorite bedding and feeding areas. Warm temperatures had melted much of the snow in the open fields and south-facing slopes. The only remnants of the harsh winter were a few skeleton snow drifts that remained on the downwind side of hedgerows. Robins were beginning to return from wintering in the South, and their morning calls of "cheerly, cheer up, cheer up, cheerly, cheer up" were a welcome sound to wildlife that had experienced winter's wrath.

As dawn broke, Buttercup slowly licked her outstretched foreleg. Between licks, the bedded doe closed her eyes to doze momentarily. Rustling leaves quickly brought her out of her sleep, and she pivoted her head to see what was coming. A ruffed grouse was walking toward her. Cautiously, the grouse walked between Buttercup and Daisy, who were bedded close to each other. Every few feet, the grouse stopped and stared at the does before continuing its march through the woods. The deer watched intently as the grouse half hopped and half flew onto a downed tree and stood motionless.

▼ If available, sumac berries are an excellent winter food source for whitetails.

▼ In very healthy deer herds, a few bucks will not cast their antlers until March. In rare cases, antler casting might occur in early April.

Within a minute, the deer were greeted to "fummp-fummp ... fump-fump-fump-fump-fump-fa-fa-fa-fa-fa-fa." Every five minutes, the grouse repeated its drum roll, oblivious to the deer bedded nearby. Though its breeding season was a month away, the male grouse was getting a head start.

After an hour, the drum rolls got the best of Buttons. The little buck stood, glanced at Buttercup and then began walking slowly toward the grouse. When Buttons got within a few feet of the log, the grouse turned and faced the fawn, as if to warn the little buck that the log was his and not to come a step closer. For several seconds, Buttons stared at the grouse, unsure of what to do next. Inquisitively, Buttons stretched his neck and head toward the grouse. The grouse aggressively flapped its wings and jumped toward Buttons, landing a few feet from the buck. It was the grouse's way of defending his drumming log. Startled, the little buck wheeled and ran back to Buttercup's side. The grouse hopped back up on the log and was soon drumming again.

▼ When March arrives, deer will migrate to south-facing slopes to feed on exposed grasses. Spring is just around the corner.

▼ By March, winter begins to give way to spring.

Vital Information

• In snow-belt regions, it's common for starvation to occur if a deer loses 30 percent of its pre-rut body weight during winter.

• If winter conditions are normal, a doe will emerge from winter in better physical shape than bucks who entered winter worn down from the rut.

• Research has revealed that whitetails do not have to go into a deep sleep to replenish cells. It's believed that if a bedded deer's heart rate drops to 72 beats per minute, it begins building cells.

• While bedded and dozing, a whitetail is constantly moving its ears. This lets them sleep and guard against danger at the same time.

• In snow country, whitetails begin returning to their spring and summer range by mid-March.

• Providing their antlers have been cast, most mature bucks begin growing antlers by the end of March.

With the morning warming, Buttercup, Daisy, Buttons and Princess rose from their beds. After a few minutes of grooming, they walked toward an open hillside, where green shoots of grass had begun to grow. They were hungry and knew the new growth would quench their appetites. Within minutes, they began to feed on the lush forage. Throughout the morning, robins hopped and fluttered between the deer as they fed. Now and then, the deer heard honking Canada geese flying over, just above the treetops. With their bellies full, Buttercup and her family bedded on the sunlit hillside.

Throughout the morning, Buttercup, Daisy and the fawns dozed and groomed themselves in the warm sunlight. On the hillside below, two deer stepped out of the woods and began feeding. As the deer fed closer, Buttercup saw it was Bucky and Skipper. After feeding, the two bucks bedded a short distance from each other, just downhill from Buttercup. Like them, the bucks had come to feed on spring's new growth. And like them, they had survived to see another spring.

CHAPTER 13

Antler Growth

Man has always been fascinated by deer antlers. Antlers are the fastest-growing bone material known, and during about 120 days from late March through early August, a mature buck can grow up to 200 inches of bone on his head. To complement Chapters 1 through 5, the following series of photos documents the antler growth process of one mature buck. I photographed this pictorial in New York state, and it's representative of other such bucks I've photographed.

April 1

There are no cookie-cutter bucks. Each is distinctly different. Some grow antlers that are narrow and tall, while others have drop tines and wide inside spreads. No two are identical, which adds to the whitetail's uniqueness. Also, the rate at which a buck's antlers grow depends on several factors, such as genetics, health, age, stress, soil quality and the overall quality of natural and agricultural habitat.

HOW IT HAPPENS

Of all the bucks I've raised, most cast their antlers from January 20 to March 10. The earliest I had a buck cast antlers was January 1 (he was injured), and the latest was April 3. When the antlers are cast, the pedicle bleeds, causing a scab to form. The scab then heals from the outside of the pedicle to the center. When fully healed, the top of the pedicle is covered with a brownish-gray skin, with a small light-gray dot in the center of the pedicle. Daylight increases as winter wears on, setting the stage for antler growth.

When day length reaches a certain point, blood begins flowing to the pedicle, and antlers begin growing. The

Velvet makes a buck's antlers appear much larger than they actually are.

skin covering the pedicles pushes upward through a series of arteries that carry blood to the antlers as they grow during the next four-plus months. During the growing process, the blood flow forms a protein base upon which minerals are deposited.

The skin covering the growing antlers is called velvet because it feels like velvet. The velvety feel is actually the result of hundreds of tiny hairs that grow out of the skin. These hairs serve a purpose: to alert the buck of danger when the antlers touch brush or other obstacles. Because the velvet consists of blood vessels, the antlers are warm to the touch and only slightly cooler than a whitetail's 101-degree body temperature. The velvet also makes a buck's antlers appear much larger than they actually are.

▼ **May 1**

▼ **June 1**

Most bucks are finished growing their antlers by August 10. Then the hardening process begins, and it usually takes 20 to 25 days. Though the antlers will be solid bone when the velvet is peeled, they are far from hard during the early stages of growth. From the time they begin growing in late March until about July 15, a buck's antlers are bulbous and pliable. Therefore, it's not uncommon for a buck to cut or even sever a beam or tine during the growth process. If the antler is badly cut or severed, the buck can bleed to death.

If a buck is mature and has everything going for him, his antlers might grow a half to one inch per day, especially from June 15 to July 15, when daylight is greatest. During this 30-day period, antler growth explodes.

TIME LINE

April: From when antlers begin growing in late March through the end of April, growth is minimal. That's mostly because the amount of daylight is much less than what it will be in June. Second, most bucks are still stressed from the long winter, so their body is in recovery mode. Another reason antler growth is slow

◀ **July 1**

▼ **August 1**: Antler growth is complete.

during April is the lack of quality food. Spring green-up doesn't normally occur in Northern states until mid- to late April. By the end of April, brow tines and one to two inches of additional antler beam should be easy to spot on a fully mature buck.

May: In most whitetail locales, May bursts onto the scene with abundant high-octane food, natural and man-created. Natural foods and forages preferred by whitetails are high in protein and other essential nutrients throughout May. This lets a whitetail's body condition improve from the stress of winter, setting the stage for rapid antler growth. By the end of May, the G-2 points (second point on a typical rack) should be noticeable. In addition, the antler beams should now be about half of their ultimate length.

June: The lyric, "Summer time, and the living is easy," is about the best way to describe what June is about for whitetails. If rainfall is normal, nutritious food is lush and readily available. Further, the amount of daylight is at its annual peak, providing the hormonal support that lets antler growth accelerate. By late June, almost all the primary points on a rack will have started to grow.

July: The month begins with a buck's antlers being very bulbous. As the month progresses, the antler's beams and points will finish growing. By July 20, a mature buck's antlers should look massive. Having adequate rainfall is critical to ensure that optimum antler growth continues during July. If a drought occurs, the nutrition level of food decreases, which can cause a decrease in antler growth. When July ends, the antlers on most bucks are fully formed, ushering in the antler-hardening process.

August: Most Northern bucks will have completed their antler growth by August 10, and the blood flow to antlers diminishes. For the next 20 to 25 days, the antlers will harden. During this time, the overall size of antlers actually appears to decrease because the velvet covering the antlers shrinks as the blood flow slows. Sometime in late August through mid-September, most bucks will peel the velvet from their antlers.

WHAT DOES IT TAKE?

Many believe that great summer growing conditions are critical to optimum antler growth. Though extremely important, there is far more to the antler-growing equation than a great growing season. For a buck to reach his potential, he must be healthy and have great food sources year-round. If either of those factors is lacking, a buck will not reach his potential.

October 1: Buck in hard antler. Boone and Crockett score is 154 typical.

Antler growth is nothing more than an extension of the animal's body condition. If a buck's body is not healthy and well maintained with proper nutrition, maximum antler growth is not possible.

During the antler-growing months, protein forages of 25 percent or more should be available to see what a buck is capable of producing. In addition, there must be adequate rainfall to ensure that forage protein levels remain high. If a drought occurs, forage protein levels decrease, and fiber content rises, making food sources less nutritious. Consequently, when droughts occur, nutrient levels required for optimal antler growth are not available in deer foods, causing antlers to be smaller.

It's vital to have great nutrition during the antler-growing season, but it's just as important to have excellent food available during September through March.

CHAPTER 14

Capturing the Moment

Nature photography has added greater meaning to my life. My quest to get one more photo has often kept me in the bush when I should have called it a day. Through nature photography, I've come to truly appreciate God's incredible creation. I firmly believe that my mission in life is to photograph the wonders of God's handiwork and share them with others. Largely, this is why I do what I do. The other part is that I love it so much. I owe my career and knowledge of nature to photography. Failing to share some tips and the story behind my photos would leave this book incomplete.

I purchased my first 35mm camera and telephoto lens while serving with the U.S. Air Force in Vietnam. I was 22 and had never owned a camera. Fortunately, I had a boss in Vietnam who was an excellent photographer, and under his tutelage, I learned a great deal about film and the art of photography. After being discharged from the Air Force in December 1970, I immediately began photographing white-tailed deer. Though capturing deer on film was my initial focus, it didn't take me long to become passionate about photographing all of nature.

Most who make their living as nature photographers say they have the best job in the world. I definitely feel this way, but that doesn't mean it's an easy profession. Believe me, it can be disheartening at times, especially if you don't know what you're doing. The essence of wildlife photography is summed up best by my good

▼ Understanding how light works is a critical component to photographing nature. For landscapes, the magic light occurs the first and last two hours of the day. My goal when I took this photo was to capture the reflected autumn hillside in the water. Waiting until the end of the day let me capture the moment.

▼ Rather than taking the photo with the sun at your back, try shooting the subject with side light. Rather than taking this photo of two bighorn rams in direct sunlight, I opted to photograph them side lit. Doing so kept the lighter colors from washing out and provided greater contrast to the mountains in the background.

▲ Turkeys can be difficult to photograph because of their dark color. Photographing them in sunlight can bring out the iridescent colors of their feathers. Also, most great turkey photographers are excellent callers. Knowing the bird and how to call them let me get this gobbler in full strut on the log.

friend and master whitetail photographer Mike Biggs, who once said, "Wildlife photography consists of a series of repeated attempts by a crazed individual to obtain impossible photos of unpredictable subjects performing unlikely behaviors under outrageous circumstances." Though humorous, this is, in a nutshell, what nature photography is about.

Because of the unpredictable nature of wildlife and lighting conditions, it's impossible to be successful every time you venture into the wilds to photograph. However, having the right equipment, knowledge of the animals, an understanding of light and composition, and a heavy dose of persistence lets you create great images of nature.

EQUIPMENT

I shot Nikon film cameras for the first 37 years of my career, but I've been using top-of-the-line Nikon digital cameras exclusively the past several years.

▲ Overcast conditions provide a different kind of magic light. It's called flat light, and it's great for wildlife images, especially when high overcast conditions persist throughout the day. I photographed this bull elk on a cold, frosty morning. The conditions let me capture the bull's steamy breath as he bugled.

▲ First light plus side light equals stunning photo conditions. Using this equation, plus having a background that was shaded from the sun's first rays, let me capture this whitetail portrait. One of the most important aspects of getting this photo was that I had scouted the location ahead of time, which let me be in the right position to get the photo if everything came together.

◀ Flat light is also great for macro photography. My goal in this photo was for the maple leaves to be in sharp focus and the background out of focus. I accomplished this by taking the photo with the lens set on its lowest "f" setting.

Though I loved shooting film, it has — for all practical purposes — gone the way of the dinosaur, at least for wildlife. So if your ultimate goal is to take high-quality photos capable of being published, shoot a digital camera greater than 8 megapixels.

Currently, I rely heavily on six Nikkor lenses; a 20-35mm f2.8 AF, 28-70mm f2.8 AF-S ED, 70-180mm f4.5-5.6 AF Micro, 70-200mm f2.8GII AF-S VR, 500mm f4 AF-S and 200-400mm f4 AF-S VR ED. These lenses are fast, extremely sharp — and expensive. They let me photograph when the light is less than adequate.

For landscape photos, the 20-35mm, 28-70mm and 70-200mm are my workhorse lenses. Most of my macro photos of wildflowers and ultra-close-up shots are taken with the 70-180 micro lens. When it comes to shooting birds and big-game animals, I rely on the 70-200, 200-400 and 500mm lenses.

To start photographing whitetails and other wildlife, a zoom lens in the 70-200 range is essential. Also, it's best to get the lowest f setting you can afford. I own a couple of 70-200 lenses that are f2.8, and they let me photograph in dim light (the smaller the f number, the

▶ Paint your photos with light. When I first walked this beach at Acadia National Park in Maine, I was struck by the size of the boulders in this stretch of coastline. Immediately, I thought of what the first rays of sunlight coming off the Atlantic would look like. The next morning, I arrived before daylight, positioned my tripod and waited for the sun to show up. Watching the way the color and contrast changed was magical. The tip is to form the picture in your mind and be there.

▼ Denali sunset. I can't say enough about the value of scouting a scenic photography location. I first visited this Alaskan location during midday and thought that it had great potential for a sunset photo. Nine hours later, this is what God handed me. Then I let my camera seal the deal.

less light required to take a picture, and the brighter the camera's viewfinder). Also, most of today's telephoto lenses offer teleconverters that are matched to the lens. A 1.4 teleconverter only loses one f stop of light and will make a 70-200 lens into a 100-280mm f4 lens. People assume most deer photos are taken with long lenses. Though most are, the 70-200mm is my workhorse lens and one of my favorites for to "shooting" whitetails.

For someone serious about photographing deer and other wildlife, a 300mm — or better a 400mm or 500mm — is a must if you want to reach hard-to-approach animals. A 300mm (6 power on a full-frame digital camera), 400mm (8 power on a full-frame digital camera, or 500mm (10 power on a full-frame digital camera) will let you fill the frame with an animal without risking spooking it. In most instances, the whitetail images on the covers of major magazines are taken with 300-600mm lenses. The downside of these lenses is their weight and cost.

"He who has the food has the wildlife" is a critical maxim when it comes to photographing wildlife. By taking advantage of a favorite whitetail food source, I set up for this early-morning sunrise. The key is to know the animal, know the location and be able to execute the photo plan.

Lighting is key to nature photography, and getting dramatic photos requires forethought and planning. The first and last hour of sunlight often provide "magic light."

Because of the nature of the sensors on many top-of-the-line digital camera bodies, the magnification of most lenses will appear to be about 1.5x greater than when they are used on camera bodies with full-frame sensors. Longer telephoto lenses can also blur out the background, which is aesthetically pleasing when doing animal portrait photography.

With one exception, I always try to shoot the lens/camera combination mounted on a very sturdy G1325 Gitzo carbon tripod, with the camera mounted on an ARCA or Wimberley tripod head. The lens I frequently shoot off a gunstock is the 70-200mm. The gunstock allows me the mobility often needed to capture action photos.

In addition, I use a full complement of lens filters, with a UV, polarizer and warming filter used most often. The only other equipment of note in my bag is a quality pop-up photo blind that lets me keep from spooking animals.

KNOW THE SUBJECT

Great wildlife photographers know the animal they are pursuing. The more you know about an animal's behavior, the better the chances of getting that once-in-a-lifetime photo. When I began photographing whitetails in 1970, there were few places you could photograph animals that were not hunted. Because of zoning regulations that prohibit hunting, there are now hundreds of locations to photograph suburban whitetails across America. Asking questions and knocking on doors is the key to learning where they are. In addition, national wildlife refuges, and state and national parks offer incredible photo opportunities for wildlife photographers.

▲ Rule of thirds. Understanding the rule of thirds is the foundation for great nature photography. The rule states that the subject image should be divided into nine equal parts by two equally spaced horizontal lines and two equally spaced vertical lines. Compositional elements should then be placed along these lines or at their intersections.

▲ Using the rule of thirds helped make this scenic of Canada's Pieto Lake possible.

▲ Shoot most scenics at an f setting of f8 or higher to attain greater depth of field. In this photo of Wyoming's Grand Tetons, I wanted the reflection and water in the center and bottom of the photo and the mountains and sky in the top third.

TIPS TO REMEMBER

When I conduct photography seminars, I emphasize to attendees that their goal should be to "make photos, not take photos." There is a vast difference. Rather than reacting spontaneously, think creatively when practicing the art of nature photography.

When I began photographing whitetails, I took photos. In the early 1970s, I was more intent on just getting a deer in the frame rather than thinking about composition, lighting or depth of field. All these aspects take time to develop, but with a little knowledge, you can shorten the learning curve.

Understanding light: Light is the key to nature photography, and when possible, I try to position the animal — or me — so they will not be in direct sunlight. If I have a choice, I like to side-light my subject. This kind of lighting makes for more dramatic photos, but getting it often requires forethought and planning. Also, the first and last hour of sunlight provide what I call "magic light" — the kind all nature photographers dream of shooting in. And don't quit if the day is overcast. You can accomplish some of the best wildlife and macro photography when the light is dull and flat.

Composition: Like lighting, the composition of a photo is essential to its appeal. When composing whitetail shots, or any wildlife and landscapes, I try to think how the subject will look best. Try placing the subject off-center in the picture so it becomes a part of the scene rather than the central focus. To enhance the photo's composition, try to find a tree or some other object to frame the subject with. Also, look for curves and diagonal lines when making scenic photos. These compositional techniques can make photos much more appealing from an artistic standpoint. To put it another way, strive to make your photos tell a story. That doesn't mean I don't like to take tight portraits, because I do. However, I try to get artistic whenever possible.

Avoid clutter: Before pressing the camera's shutter, look for clutter and bright or dark spots in the

When possible, show scale in your photos. When I took this photo of an Alaskan moose, I wanted to show the awesomeness of Alaska by capturing the moose and the scenic in one photo.

viewfinder that can detract from the photo. Your goal should be to make the photo as clean and desirable as possible.

Horizontal or vertical: It isn't always easy, but strive to determine whether your scene will work better as a horizontal or vertical photo. If one of your goals is to submit photos to magazines, you'll want to think about composing the photo in vertical and horizontal formats.

Show scale: If possible, attempt to place an animal or person in your scenic photos. Doing so will show the size and scale of what you are photographing.

Focus on eyes: When taking portraits of animals, always focus on the subject's eyes. The eye is the center of attention and reveals the soul and character of the subject. In addition, the glint of the eye adds life to the photo. I also like to take pictures from the subject's eye level or lower. If the subject is a fawn on the forest floor, I photograph on my belly.

Depth of field: For animal portraiture, shoot with a shallow depth of field (with the lenses' aperture wide open). Doing so will blur out the foreground and background, giving the photo a clean look. For most scenic photos, I take them with the lens set on f8 or higher. This renders the foreground and background in sharp focus.

Using the rule of thirds, let me tell of the whitetail's struggle in winter.

▲ If the situation warrants, try to frame the subject. In this case, I framed the Grand Teton Mountains with the foreground and an ancient tree.

◄ To take advantage of changing light conditions, I framed the bright light of this slot canyon with darker contrasting rock formations.

Nature photographers are not born, they are made. Scan books, magazines and websites to see how others are photographing wildlife and nature.

Early-morning light, a deep blue sky and the diagonal flow of the river helped make this photo of the Canadian Rockies possible.

Stopping the action requires a shutter speed of at least 1,000 of a second. My camera's shutter speed for this photo was 1,250 of a second at an ISO of 1,250. This is an example of one of the beauties of digital cameras, because I could never have captured this image shooting film.

Stop the action: Perhaps the greatest challenge in nature photography is capturing action. Things can occur quickly in the wild, and getting it right doesn't just happen. To stop fast movement, you must shoot a shutter speed of at least a 500th of a second or a 1,000th of a second or higher if you have enough light.

Blur the action: To add motion to the photo, slow down your camera's shutter speed. You can create a sense of motion by shooting at a shutter speed of 1/15 of a second or slower. Letting the camera's shutter remain open for a second or more will make the water appear to be moving.

Keep it level: When taking scenic photos, make sure the horizon is level in the camera frame.

Break the rules: Don't take all your photos from eye level. Try new perspectives, such as shooting into the sun or taking photos from ground level. In short, try to be different.

Dare to learn: Nature photographers are not born. They are made. Scan books, magazines and websites to see how other photographers are photographing wildlife and nature. Read all you can about how certain equipment — lenses, filters and software — can im-

Stopping the action can be tricky. In this case, I was aware of the ruffed grouse's drumming, so I waited for the pause in his drumming sequence to stop his wings.

prove your work. The day you stop learning is the day your dream begins to die.

Be there: Ultimately, great photos cannot be taken without being there. As mentioned at the beginning of this chapter, wildlife, lighting conditions and weather can be unpredictable, so capturing a great photo requires time and persistence. I learned a long time ago that photographing nature is an incredible challenge that requires knowledge and the will to keep shooting when others have quit.

Showing movement in photos can create a special mood. Capturing the water's flow requires a sturdy tripod and very slow shutter speed. I took this photo using a 20-35mm lens with a variable neutral density filter. The camera's shutter speed was four seconds at ISO 200. The lens was set at f13.

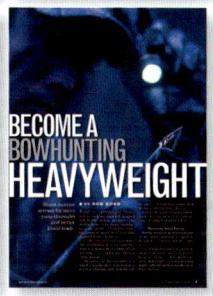